PERSONAL FINANCE FOR TEENS

EASY MONEY SKILLS TO BUILD WEALTH, CREATE
FINANCIAL FREEDOM, AND LIVE LIFE ON YOUR OWN
TERMS

B. B. GLYNN

CONTENTS

INTRODUCTION

Hello, future financial expert! Welcome to my book, which I've crafted just for you to help you achieve your dreams and fulfill your life's goals.

I have a question for you: Can you imagine owning, as a teenage investor akin to Warren Buffet, the world's wealthiest individual, your own basket of investments? Sounds like a dream carved out of a DC Universe franchise, right? I mean, you aren't Batman, are you?

What if I told you that with the right book and the proper guidance, you can set yourself up for success at an early age? "Is that possible, and what do I need?" you might ask. It's actually quite simple! All you need is financial literacy! You see, financial literacy leads to financial freedom, and that's why I'm here —to guide you on your journey to financial independence. This journey is crucial for your development into adulthood and living a satisfying life.

The world has changed a lot, and now, presents new challenges, perhaps even more challenging than those of my time. However, one thing remains constant: With adulthood comes great responsibility. You've already recognized this, and that's likely why you've acquired this book. Schools don't teach us how to handle money, and parents often lack the necessary knowledge themselves. Yet, here you are, about to waltz through the symphony of life, unprepared and left to navigate the increasingly complex world of finance.

I'm certain you've lain on your bed, pondering:

"Which job suits me best?"

"Is credit score my new BFF?"

"Do I always have to pay taxes?"

"Can I conquer the finance game?"

Deep breath—no more worries, my friend. The very purpose of penning this book is to aid individuals, like you, who grapple with and lack a solid grasp of the world of finance. This is my way of giving back to society—assisting younger individuals who remind me of my past self in overcoming the challenges I faced at their age. I aim to take your hand and steer you on this journey! Consider me a guide and a friend, and together, we shall conquer the world.

THE SUPER POWERS I'M GOING TO SHARE WITH YOU

Buckle up, kiddo! You're about to embark on a journey through essential tips that will shape your life for the better. Keep that curiosity alive, as it defines your teenage spirit. Absorb the wisdom within these pages, and I assure you—your future will shine brightly! Without further ado, let's dive right into the exciting content of this book.

Streamlined Financial Education

I'm sure you've come across an endless list of books and materials on finance. These resources are scattered everywhere. However, you're in luck: I'm genuinely invested in your success!

This is precisely why I've compiled and condensed all the valuable information into a stage-by-stage guide, much like the progression of levels in an adventure video game. Starting from level 1 and moving upwards seems like a logical approach, don't you think? Not only will the information be more digestible, but you'll also save a substantial amount of time.

Quick Start to Investing

Once you get a grip on your personal finance, you can take on more exciting adventures. Have you ever danced to the tune of "investments?" Picture it as planting seeds, and nurturing patiently till they burst into towering trees, yielding a bounty of fruits just for you! Throughout the book, you'll learn the in-

and-outs of investment, the thrilling and not-so thrilling aspects of it and uncover the powerful potential for building wealth. You don't want to miss the investment trends!

Guided Path to Credit Mastery

I'll unravel the mystery behind credit scores, credit cards, and smart borrowing. It'll serve as your road map, so to speak, to understanding the secrets of credit while offering clear steps to construct a positive credit history without falling into tricky traps.

Entrepreneurship and Job-Hunting Tips

Making money requires a job, right? So, how do you land one? It all boils down to your knack for showcasing your skills or products (if you're business-savvy). This book is your ultimate toolkit for crafting killer resumes, nailing interviews, and even kickstarting your own business ventures. It's like having a secret stash of strategies that supercharge your money-making game, all while you're still in your teenage glory. So, are you ready to show the world your potential? Are you ready to unlock your earning power? If yes, then you've come to the right place!

Smart Shopping Skills

Ever strolled out of the mall, wondering where all your money vanished? Tag along with my "mom's shopping" guide to becoming a savvy shopper extraordinaire! You'll learn how to spot marketing tricks that charm you into excessive spending. How about mastering the art of hunting down the sweetest deals—buying the same product for less? Sounds too good to be true? Not quite! If you're ready to level up your shopping game, then you know what to do!

College and Career Planning Pointers

Packing your bags for college? Wait a minute because I've got some valuable insight that'll come in handy! For instance, did you know you could set up a saving plan for college? But wait! What if I told you that you could also make a plan to successfully decode and obtain financial aid? I've got it all covered! Plus, I've sprinkled in some extra juicy tips on choosing the right career path. Don't miss out on this!

Emergency Preparedness and Insurance Know-How

Nothing beats the confidence of having a backup plan. Every financial guru worth their salt knows that the best plan is one with a backup. In the realm of finance, there are several ways you can set up your "plan B." In this book, we'd discuss two such ways: Stashing up funds in a super-secret treasure box and safeguarding your property with insurance against unexpected financial setbacks.

Tools for Long-Term Success

Feel free to adopt the techniques discussed in this book to maneuver your way through life's twists and turns while keeping your finances in good shape. Remember, life is like a game; you need to unlock the secrets to managing money and use them to level up. Ultimately, you want to ensure your golden age of retirement promises nothing short of vacations, fishing trips, dinners, and watching the marvels of the world unravel before your eyes. Get ready to build a rock-solid foundation for your finances and secure a radiant future for yourself and your family!

STILL NOT CONVINCED?

Ever wondered what sets people like Bill Gates, Warren Buffet, or Elon Musk apart from others? I assure you, it's not luck or magic. A great deal of their success rests on the power of personal finance! If I could, I would climb up to the Eiffel Tower and scream at the top of my lungs, "Listen, teens, fix your personal finance!"

Thankfully, the demand for financial knowledge is on the rise. More and more people are calling for personal finance education to be added to school curricula. In fact, that number sits at around 63% of American adults (Ward, 2019). You don't need to wait for those changes to take effect. The financial knowledge you need is right here sitting in your hands!

1

UNDERSTANDING THE BASICS OF MONEY MANAGEMENT

E very parent thought they'd leave behind a huge pile of wealth for their kids—instead, Gen X has left a huge pile of financial burden. The good thing is, Gen Z are woke to the financial woes of their predecessors and are doing everything possible to not make the same mistakes. This is apparent because 93% of teenagers have hit the stage where they are asking for some financial education. In fact, only 45% of adults wished they'd start up some business, compared to 54% of teens who are aiming to be entrepreneurs (Tenny, 2022). If you're reading this, well done!

My dear, together we're aiming to change the narrative and the sad story of bad finances you've inherited from your parents. But, to do that, we must begin with the fundamentals of finances; that aspect of yourself that controls how you see the world in terms of finances and money. That is none other than your mindset.

MONEY MINDSET MATTERS

When you hear the word "mindset," what comes to mind? Your beliefs? Perceptions of reality? Some religious mumbo jumbo?

Your mindset is the lens through which you view the world. For example, if you want to be a skating champion, you must think and act like a skating champion. Many think this "mindset" concept is just some invention coined by motivational speakers to rob people of their money. On the contrary, it is something as real as the dawn of day!

Now, a mindset can build you or destroy you; it totally depends on what kind of mindset you choose to adopt. Some of us have inherited a negative mindset from our parents and environment, which continues to affect our daily lives. But don't give up yet. Amazingly, mindsets are like bread dough; with the proper technique, you can mold it into something delicious and fluffy!

Do you think everyone can learn to play the piano? Yes, some people may catch on quicker than others, but everyone can learn to play. Who do you think would become a master at playing the piano? The student who looks in the mirror and tells themselves, "I can't learn the piano, I'm too shy," or the student who looks and goes, "Hey, buddy, it's not gonna be an easy ride, but you've got this." The second student? Right! That's the mindset for you!

Overcoming Common Money Myths and Misconceptions

Allow me to task your wonderful imagination once more. Picture your favorite social media app. Just as you curate your social media feed with things that interest, inspire, and uplift you, your mindset also curates your thoughts. Here's a little tip not so many know about: Your mindset governs your perception of life and can influence your actions. That's why it's crucial that you fill it with positivity, motivation, and dreams to create a feed of success.

Another nugget of knowledge for you! Your mindset greatly impacts your finances. Here's one of my favorite quotes by Napoleon Hill:

 "There are no limitations to the mind except those we acknowledge. Both poverty and riches are the offsprings of thought."

Have you heard of any of the following sayings?

"Money is the root of all evil."

"Money cannot buy happiness."

Yes? I guessed so. These are unhealthy perceptions about money that weigh people down from emerging from the seas of financial woes. What mindset do you have about money? I want you to take a deep breath, close your eyes, and access your thoughts. Have you ever told someone:

"Money changes people."

"Money is scarce."

"Money causes conflict."

"All rich people are in the Illuminati." Oh, this one made me laugh a lot!

"It is better to be poor and happy than rich and miserable."

If you do, I don't want you to feel bad. We were all raised on such negative mindsets. But I've got good news for you! The mind is like malleable metal; it can be bent whichever way you wish!

The first step to reshaping your mind is to become aware of the thoughts drifting through your mind. Next, you want to audit your mindset about money and counter it with positive affirmations like, "I can shape my financial future," or "I'm in control of my money decisions."

CULTIVATING A POSITIVE MONEY MINDSET

Mindset doesn't end with piano lessons; it can equally determine if you're going to be successful in life or not! Now that you know what mindset is, you also have to know why it's important to shape your mindset. Here are some incredible ways to shift your mindset from a negative one to a positive one.

Envision your epic quest: This is where you ask yourself the important questions.

"What do I expect out of life?"

"How much do I need to earn?"

"Where do I see my finances in one year, three years, 10 years, or retirement?"

"When I get old, do I retire completely or start a side hustle?"

"What destinations would I like to visit?"

Frequently ask yourself these questions. Just like having your favorite playlist on Spotify, keeping track of your life goals and revisiting them often is like a personal pep talk! Mind you, you want to set goals that are totally doable.

Craft your short-term financial roadmap: Divide your goals into two subcategories: Short-term goals and long-term goals. Always begin with short-term money goals before mapping your long-term goals. Think of short-term goals like baby steps that make your money goals easier to achieve. I mean, it's definitely easier to make $20 per hour than it is to make $10,000 per month. Plus, making $10,000 sounds a lot more difficult—too difficult that the mind shivers when it sees such figures. But, if you break that $10,000 into smaller chunks of $20 within a certain time frame, you'd feel positive that that goal is within reach.

Ignite your motivation rockets: Always be on the lookout for inspiration. Sometimes, life gets cloudy, and it seems like the end is near. In such times, finding some motivation juice to ignite your spirit and getting back in the race will prove crucial.

Where do you get such inspiration? A singer would listen to inspiring music, right? That is how it works for finances too.

Thanks to social media, it's so easy to find inspiration. If Elon Musk inspires you, follow him on X (Formerly called "Twitter"). It could be anyone! It could be a blog about personal finances! You just need to make sure the burning flames of growth and success are never quenched!

Join the positive money mindset tribe: Nothing pushes you to do better than having a bunch of friends who share the same dream! If you have such people around you, then you want to keep them close. They'd be there to rub your shoulders when things go south or tap your back during times of triumph!

Ditch people with a negative mindset: Are there people who always complain of the ill deeds of money? Does anyone in your circle think money is a tool for evil? These are people with negative mindsets, and you shouldn't mingle with such people. The old saying, "Show me your friends, and I shall show you who you are," stands true in this case.

Record any thoughts about money: Everyone needs a little motivation to keep going. Keeping a journal isn't a new phenomenon to you, is it? Reading a journal is almost like re-living the moment! It might make you smile or even cry.

What if I told you that journaling could be used to motivate you? Yep! Writing down your financial goals can juice up your motivation to take action. It could serve as your armor against negativity. Who would have thought that a journal could hold so much power?

Get some financial knowledge: The only way to level up your finance game is to invest in financial education. But you're doing that already! Don't give up though; continue to explore and learn to unleash your inner money guru. This reminds me of a scene from Game of Thrones where wise Little Finger threatened Queen Cersei with the words, "Knowledge is power." In the finance world, with knowledge comes empowerment!

Master the money puzzle with your ultimate budget: As I mentioned earlier, people are devastated by their inability to track their expenses. It's almost like their money disappears from their pocket. This ends when you become more finance-savvy. Budgeting and tracking your money gives you a broader view of your financial situation and guides you out of the money maze!

Spread joy—contribute to charity: Nothing trumps being able to shine your light on the world! Financial power isn't just a game of numbers—it's also about impact. Mark Zuckerberg, Bill Gates, Selena Gomez, Eminem, and Beyonce are all philanthropists who've shared their success with the world. Who knows? Tomorrow it could be you!

Setting Clear Financial Goals and Creating a Vision for the Future

You are an adventurer in the world of finance, and to navigate your way around it, you must plot a roadmap. This roadmap not only fuels your motivation but also steers you through life's twists and turns, ensuring that every step aligns with your desired destination. Follow these tips to create your very own road map.

Make your goal specific: Crafting clear and specific financial goals is key. Define what you want to achieve, whether it's saving for a car, college, or a dream gadget. Getting detailed brings your goals to life and sparks your inner fire!

Make your goal measurable: Instead of vague wishes like "save money," say "save $500 for a new bike." Measurable goals offer a sense of accomplishment as you watch the numbers grow.

Give yourself a deadline: Time to put the "go" in "goal!" Set a realistic deadline to create a sense of urgency and structure. A timeline helps you prioritize and stay committed. It's like having your own countdown to success pushing you to make every day count!

Make sure they're your own goals: Your goals, your dreams, your journey! Don't chase someone else's dreams. Tailor your goals to align with your passions and values. You'd find that the personal connection boosts determination and enthusiasm.

Write your goals down: Grab a pen and let your dreams flow onto paper. Putting them in writing seals the deal—it solidifies your commitment and increases the chance of achievement. Time to embrace the magic of written dreams!

BUDGETING MADE FUN!

Hey! Budgeting, at first glance, may seem like a daunting task. I mean, it's no walk in the park to keep track of your spending. I sometimes let myself lose. Fifty-five percent of Americans don't use a budget even though they are well aware of its purpose (Dow, 2021). Most people just don't feel like penning their finances down, and we can't blame them.

Thanks to technology, you could ease your way into becoming a budgeting sorcerer! You might not have guessed it, but you could learn how to budget while having fun. I play games like Monopoly all the time—who'd have thought it could help shape your budgeting skills? Let's jump into some games that could help you become finance-savvy!

The Best Budgeting Games

To help you accomplish this feat, I've compiled a super-fun list of games that will teach you how to manage money while having fun. Good luck!

Monopoly: Monopoly is a thrilling game about buying, selling, and trading properties. A winner is declared when they've acquired the most resources and their opponents are out of money. Monopoly features many elements of personal finance; you learn to budget wisely, make investment decisions, and deal with unexpected expenses. Fun fact: It was originally designed by Elizabeth Magie in the early 1900s and called "The Landlord's Game."

Modern Art: Modern Art brings a creative twist on personal finance learning. Instead of trading properties, you'll be trading and investing in valuable art pieces. Ever seen auctions in the movies for paintings like the Mona Lisa? You'll be playing the role of a museum looking to trade and invest in valuable art pieces.

For Sale: For Sale is a fast-paced world of property dealings—a thrilling adventure that hones your financial skills. The game operates very much like Monopoly: Your goal is to buy properties cheaply and flip them for maximum profit. For Sale gives you a taste of real estate investments–making smart moves is your only ticket to success.

The Game of Life: The Game of Life is my top recommendation because it's not just about playing–this game brings real-world money decisions to life! Like it's said in the name, it features basically all aspects of life that have to do with finances. Can you imagine having to simulate life from the stages of childhood, to getting an education, choosing a career, getting married, and even having kids?

Other Honorable Mentions: Some other games that are equally awesome, and I think you should try out, include: Cashflow, Animal Crossing, Minecraft, Money Island, Financial Football, and Stardew Valley.

P.S.: Most of the games mentioned do have their physical board and online versions. For example, you could get Monopoly for free on your Apple App Store or your Google Play Store.

Tips for Playing and Winning Budgeting Games

To quote Little Finger again, "Knowledge is power." To win any game, finances, or aspect of life for that matter, accumulating knowledge is key!

Pick age-appropriate games: Don't go playing games meant for a five-year-old. The games you pick should match your skill level and age. If you're new to these games, start with simpler ones, then progress to more challenging ones.

Master the game rules: Like any other game, start by understanding the rules of your budgeting game. Get a good grasp of how the game works, its objectives, and how money flows.

Formulate strategies: No warrior survives the battlefield without a careful, thorough strategy. For instance, playing online versions of the game will connect you with veterans who've spent hours, days, and even years formulating a strategy. Never enter the battlefield without a course of action, soldier. Over and out!

Make it fun and engaging: Playing a game that's slow-paced quickly becomes boring. I guess, compared to watching the infinite reels on Instagram, Monopoly doesn't stand a chance in offering the same level of excitement. It's up to you to decide to keep the game engaging. Build new challenges for yourself, compete with family and friends, and join online communities with shared interests.

Reflect and discuss: After a battle, go back to the drawing board. You're a master planner, remember? Take some time to analyze what led to your victory or excruciating defeat. What smart or reckless choices did you make? Spent too much money on a property? Got tricked by your opponent? Your journal should be your personal advisor, helping you to see what you missed. Don't shy away from bringing family or friends to the party!

Stay positive: Keep the flames burning! Don't beat yourself up if you're not winning at first; remember, every skill takes time to master. Celebrate your wins, learn from your mistakes, and level up your money management skills!

BECOME AN EXPENSE TRACKER

Grab a pen and paper because things are about to get technical! Let's dive into some hot tips for creating a budget that suits your lifestyle.

The first step to creating a budget is to picture what you really want out of life. On a sheet of paper or your notepad, jot down what you hope to accomplish in life. Until you can set up an action plan to fulfill your dreams, they will remain what they are: Dreams.

Step two: Create a detailed list of everything you spend money on—from ice cream to tuition fees, track it down. Go on ahead and do that. I like to group my expenses into categories. For instance, you could group your expenses into food, clothing, entertainment, utilities, and so on.

Here's an expense tracker sample to begin with:

Monthly expenses	
Categories	Price ($)
A. Food	
1. Bread	
2. Milk	
3. Eggs	
Total:	
B. Entertainment	
1. Apple Music	
2. HBO monthly sub	
3. Netflix	
4. School club monthly fee	
Total:	
C. School supplies and expenses	
1. Library fee	
Total:	
D. Utilities	
1. Electricity bill	
2. Water bill	
3. Internet service	
Total:	
E. Savings	
1. Savings for emergency	
2. Savings for college	
Total:	

The next question to answer is, how much do you earn monthly? Prepare a table or spreadsheet that outlines all your sources of income—record every penny that goes into your pocket; it doesn't matter the source.

Monthly Income	
Source of Income	Amount ($)
1. Dad/Mum monthly pocket money	
2. Salary	
3. Summer jobs (Babysitting)	
Total	

Okay, congratulations! You've just formed your own budget. Way to go! We've successfully tracked all your expenses. Now, we need to answer the important question: "How do I know how much to spend?" To build a budget, we need to employ a budgeting strategy. A budgeting strategy is what tells you how much of your salary to allocate to food, entertainment, utilities, and so on. There are many ways to build a budget, but we'll talk about the:

- zero-based budgeting
- pay yourself method

Zero-based Budgeting Method

Okay, so you know how when you plan your budget for the month, you just kind of copy and paste your expenses from the last month? Well, that's not how zero-based budgeting works. With this method, you have to analyze every expense and figure out if such expenses are worth your money. The zero-based budgeting helps you answer the question, "Do I really need those Nike sneakers?" One thing I love about the zero-based budgeting method is that it ensures every penny of your income has a specific purpose.

But wait! It doesn't mean you'd be spending all that hard-earned money! The thing is money without a purpose is easy to lose. If you're not spending it, then it means its purpose is to build a savings or emergency fund.

That's enough talking; let's jump into a sample!

Expense category	Allocated funds ($)
Entertainment	
Utilities	
Pet food	
Phone bills	
Fun money	
Dining out	
Total	
Income - Total	

Normally, you'd begin with, "Oh, what did I spend last month?" We don't do that with zero-based budgeting. So, you should ask yourself, "What do I truly need for the month?"

Suppose your monthly income is $2,000 after taxes. Then:

- Figure out your monthly income.
- Divide your expenses into categories and figure out how much you need for living expenses, emergency savings, retirement contributions, debts, entertainment, and so on.
- Now, subtract your income from your expenses. You know you've done it right when your "Income" minus "Expense" return is zero.

Pay Yourself First Method

"Pay yourself first." If you're a fan of YouTube finance gurus, you've heard the phrase thrown around quite a lot. It's a simple budgeting method that ensures you've stacked up some savings before any other expense comes in. Let me explain it another way.

Suppose you have a big bag of candy. You want to make sure you don't eat it all at once, so you decide to give yourself a certain number of pieces each day. You might set aside five pieces of candy for each day of the week. That way, you can enjoy your candy without running out too quickly.

The pay yourself first method works the same way, but with money instead of candy. You decide on a certain amount of money that you want to save each month, and you treat it like an allowance. Just like you wouldn't want to run out of candy too quickly, you don't want to run out of money before your next paycheck.

Put Your Budget in Your Pocket

Today, you have something I didn't have when I was your age. Something so powerful it's like a hundred times better than computers of the 1900s. That thing is none other than your mobile phone. Yep! You take your pictures, videos, and music with you, don't you? Why not your budget? There are very useful budgeting and saving apps on both the Apple App Store and the Google Play Store. Here are a few you could look up.

- You Need a Budget
- Mint
- Simpli by Quicken
- PocketGuard
- EveryDollar

P.S.: Some apps allow you to perform basic budgeting for free but offer more sophisticated finance services at a monthly fee.

What I Want Versus What I Need

Go over your expenses again. Take a look at the items you've listed. I want you to put an asterisk beside the most essential things you need to operate. Now, for my magic trick; allow me to guess the asterisked items. Did you have the light bill? Yes? Aha! Trick one! How about the water bill? Bingo! Trick two! Okay, a last try. Is transportation on the list? I told you I was a magician! These are necessities for living, right?

What am I driving here? The asterisked items are essentials to your livelihood and are called "needs." You obviously can't do without them, while those not asterisked items are stuff we desire but are not essential for our survival and are called "wants."

Communication is important; therefore, we *need* a mobile phone. But you *don't need* the latest iPhone 14—you *want* it. That's because you already have a very great model, and your desire to join the crowd makes you think you need another new iPhone. A quick glance at the statistics and you discover that 8.9 million Americans go into debt to buy new tech. Guess what? Sixty-nine percent of those folks spent most of their loans on mobile phones (Davis, 2023)!

Am I saying you shouldn't buy an iPhone? Of course not. I want you to understand the consequence of spending too much on wants. Wants would leave a huge hole in your finances!

THE POWER OF SAVING

Saving isn't just about being prepared for emergencies–it's also about grabbing hold of your dreams and making them real. Want to travel the world, start your own business, or sail through college without drowning in debt? Saving creates a golden path to making those wild dreams of yours a reality.

Really, what's plan A without a plan B? Stashing some money aside creates a feeling of security—that you have something to fall on if things go south. Believe me; things often do go south more than we'd expect.

This is no walk in the park either: If you don't start now, you'll find it quite difficult to start later. Starting today enforces the habit—the habit of building a solid life of financial wonders. Did I mention savings have the ability to grow? More on that later!

Saving With a Purpose: Short-term and Long-term Savings

Good! You've realized the importance of saving. But you need to answer the question: "What am I saving for?" Are you saving to buy a new gadget, college, vacation, or retirement? You'll notice that some life goals are more future-facing than others —for example, it takes most of your life to save for retirement compared to saving for a few months for a new gadget. In this case, saving for a gadget is referred to as short-term saving, while saving for retirement is, you guessed it, long-term saving.

How do you know what is classified as long-term and short-term respectively? That's easy! Simply calculate the time it takes to save for such a goal. What goals do you have? Identify whether they are short-term or long-term goals. If it'll take you a few weeks to about a year to save, then it's a short-term goal. Otherwise, classify it as a long-term goal.

Strategies to Grow Savings Effectively

As I said before, savings offer you financial security; without them, you'd end up like 85% of Gen Z who reported that they wouldn't be able to afford a single month's worth of living expenses if they suddenly lost their jobs (ADAMCZYK, 2023). Take a look at the numbers again; 85%! What about you? Why haven't you saved money? Whatever the reason, I've got you covered. Here are some strategies that can help boost your savings game:

1. Make saving a breeze! Set up automatic transfers from your paycheck to your savings. Out of sight, out of mind, and your savings stack up without effort.
2. Life's curveballs are no match for you. Create an emergency fund to handle surprises like a champ; a safety net for peace of mind.
3. Crush high-interest debt like a superhero. By tackling it first, you save big in the long run. Your wallet will thank you!
4. Dream big, save smart. Whether it's a new gadget, college, or that epic trip, allocate funds to specific goals. Watch your dreams turn into reality.
5. Organize your savings game! Use separate accounts for different goals. It's like having pockets for every mission, making your financial life super organized.

END OF SESSION

Financial woes have been passed down through generations; Generation Z is emerging as a force to reckon with. Together, we've discovered some valuable insights that you can apply to transform your financial situation.

We began with the cornerstone of your financial mindset, which is the lens through which you perceive the world of money. It can take you to the top of the skyscraper or derail your financial goals. Therefore, it's essential to understand the importance of your mindset, debunk myths and misconceptions, and build a solid financial foundation from scratch.

Now, the road gets a little bumpy. We often wonder about the world with vague ideas of our financial future. The second step to climbing the ladder of success is setting down realistic financial and life goals; they could be short-term, long-term, or a combination of both.

Next, put your plan in motion. Begin by eliminating all the financial black holes zapping your hard-earned money to unknown dimensions. How? Craft a budget that aligns with your dreams; monitor it, analyze it, and re-strategize often.

Get ready to gain a deeper understanding of finance and embark on a journey that will reshape your financial future!

2

MAKING MONEY WORK FOR YOU —INVESTING BASICS

According to the Bloomberg Billionaires Index, did you know that the fortunes of billionaires jumped by $852 billion from January 2023 to June 2023? Do you know what this means? The first 500 billionaires made a whopping $14 million per day (Massa & Witzig, 2023)! As a matter of fact, other teenagers like yourself have turned their lives around by investing.

In 2021, 19-year-old Adam Mlamali, from Buckinghamshire, England, transformed 200 pounds into a 200,000 pounds empire (Gayle, 2021). Was he a child to a wealthy parent? No! Did he have a network of friends affiliated with big companies? No! How, then, did he perform such a miraculous feat? Mlamali learned the ins-and-outs of building wealth from YouTube. You heard that right!

Ever wondered how wealthy people take pennies and transform them into wealth? The secret might be simpler than you think and is rooted in understanding the world of investments. I'm here to tell you that you too could become wealthy by investing. Are you ready to ignite your money's potential and let it work for you? Keep reading!

Introduction to The World of Investments

Once you crack the code of budgeting, you suddenly realize that something magical is happening. Your savings are growing, steadily piling up like an avalanche rolling down a hill. And then, a thought pops into your head: "What do I do with all this money?"

Hold on! There's something you must know. Saving and investment are completely different concepts. Imagine you have a secret locker where everything you throw in it grows in size? That's investment!

Investing simply refers to putting your money to work for you. It's all about smartly choosing where to put your hard–earned cash so it can grow over time. Pay close attention, we're about to dive deeper into the finance ocean. The money you throw into your magical locker is your *investment*. In the finance world, we call the magical locker an *investment vehicle* or *assets* or *securities*. Assets, securities, or investment vehicles are things of value with the potential to either become *more valuable* or *less valuable*. Now, there are several types of investment vehicles like stocks, bonds, real estate, mutual funds, and so on. Maybe you've bumped into one of them before; any of them sound familiar? All these work a different kind of magic on your money and also determine how much wealth you accumulate over time.

Investment is a serious game. It has its good sides and its bad sides. The good side is you get more money, but on the bad side, there's always the risk of losing your money. How much you stand to gain or lose depends on the investment vehicle or asset. Investment vehicles or assets that perform the greatest magic of them all come with huge risk, and those that perform the least come with little risk. Get the drift?

There are two ways you can go about investing; you can do it yourself like Mlamali or you could pay someone to do it for you. What's the difference? You see, you need to study a lot to become a profitable investor—like a lot!

Your job, coupled with other endeavors, might not allow this. In such a case, you have no option but to place your money in the hands of professionals. Ever heard of the "Wolves of Wall Street?" Yep! Those guys.

But we'll save that for later. Let's talk more about different types of investment vehicles that could change your life for the better.

1. High-yield Savings Accounts

A high-yield savings account is a savings account with extra perks! You see, when you open a savings bank account, your bank likes to reward you every year for saving your money with them. That reward you earn is called *interest* and is the extra cash that gets paid into your account. Interest is your greatest friend in the financial world. Why? You'll find out soon. So, with a high-yield savings account, your interest is higher compared to what you get from a regular bank account. That's it! Easy-peasy!

2. Certificates of Deposit (CDs)

CDs are another type of savings account. Unlike a regular savings account, the money you put in a CDs account must not be touched. Nope! You can't access it for a certain period, referred to as the "term." Although it sounds silly to keep your money in a bank and not be able to touch it, the banks make it up to you. They reward you with even higher interest rates because you didn't touch your money. Awesome, right? Getting paid for simply keeping your money with a bank!

3. Stocks

Let's get a little practical. What brand of mobile phone do you prefer? Apple, Samsung, Motorola, LG? My guess is it's Apple. Did you know you could own a part of Apple? Yep! To own a portion of Apple, you need to buy something called "shares." Stocks or shares represent ownership in a company.

So, buying Apple stocks gives you the right to ownership of the company. Thinking of becoming the next Apple boss? Hold your horses, kiddo! Let's try this approach. Apple Inc. is a pizza with eight slices. If you bought two slices and I bought six slices, it means I get to call the shots. In other words, the more shares you have, the more power you have in the company. Walk around the house, find the brands you commonly use, and look them up on google.

4. Bonds

A bond is a type of asset that tells the world of an agreement made between an investor and a borrower. For instance, your high school buddy, Tim, asks you for $30 and promises to pay you $1 every week for one month and then return $30 by the end of the 30th of August. You could tap yourselves on the shoulder to signify an agreement, or you could write your agreement on paper, which would include details about the loan. It could read:

I, [your name], am lending $30 to my friend, Tim. Tim promises to pay $1 every week until the 30th of August 2024. Also, on the 30th of August 2024, Tim is to repay the $30 he loaned from me.

Signed
Tim

Signed
Me

This agreement is the bond. Some important stuff to know: The $30 loan is called the "capital" or "principal." The $1 payment every week is called the "periodic interest payment." By the end of August, you'd have received a sum of $34; $4 ($1 per week), and the initial loan amount. Cool, right?

In reality, the borrower is mostly a corporation, government, or other organization. So, when you invest in bonds, you are expecting to receive periodic interest payments, plus your capital when the agreement expires. There are many types of bonds, government bonds, corporate bonds, municipal bonds, and so on.

5. Mutual Funds

Mutual funds are like the avengers, only this time, instead of superheroes, you get a collection of different assets coming together to form the mutual funds. Finance experts "collect" money (the correct term is "pool") from you and other investors and invest that money into different kinds of a collection of assets. A collection of assets in the finance world is called a "portfolio". That's another word you can add to your financial library! We'll be using it later.

6. Money Market Funds

These are a type of mutual fund where financial institutions (companies that deal with financial matters; for example, banks) gather money from you and other people. The money collected is then invested in assets like government bonds, stocks, and other assets. What makes the money market unique from other types of mutual funds is that they are targeted for short-term and low-risk assets. Assets come with different levels of risk, remember? Money market funds are designed to be a lot less risky than others while promising decent interest.

7. Exchange-traded Funds (ETFs)

Exchange–traded funds is a mix of different things you can invest in, like stocks, bonds, or commodities like gold or oil. They are a special kind of asset that sort of mimic the performance of other assets like stocks, bonds, and so on. There may come a time when you wish to invest, but you're not sure where to start. That's where ETFs come in: You'd be able to spread your investment across multiple assets, reducing your chances of losing your money!

8. Index Funds

Let's say you're a DJ at a party with a vast audience. You'd need to provide a wide range of music genres to play, right? Exactly! So, to keep everyone entertained, you build the ultimate playlist with a mix of songs from different genres that everyone loves. That way, you ensure everyone gets a piece of entertainment. Index funds are like a playlist of assets (mostly stocks and bonds) all combined together and acting as one.

Index funds often follow what's called the *stock market indices*. The stock market indices are like a curated playlist themselves, but instead of music, they represent a group of top-performing companies. A common example of stock market indices is the S&P 500. When you invest in index funds, you're getting access to all the companies, just like enjoying a mix of music genres at a party.

9. Real Estate

You've probably heard this thrown around a lot. Real estate refers to any permanent structure, like land, buildings, houses, and so on that you can buy, sell, or use for other purposes. Your parents might be real estate owners; are they living in their own house?

When it comes to investment, we're more interested in the buying and selling side of things. That brings us to real estate investments, which is real estate that puts money in your pocket! For example, hotels, motels, and Airbnb are all retail investments. Why? Because they fetch money. This might not be an option for you at the moment, but I thought you should know.

10. Alternative Investments and Cryptocurrencies

There are tons of other investment vehicles out there; for example, commodities are another common investment. What are commodities? It's like investing in things you can touch and use in your everyday life. For example, did you know you could buy gold or silver bars? Well, you aren't going to be holding the real stuff, but you could still trade them. You could also trade commodities like agricultural products and even livestock. The principle is simple; buy them when they are cheap and sell them when prices shoot up!

I'm adding another weird one to the list, and that is cryptocurrencies. If you haven't heard of cryptocurrencies, then how about Bitcoin? Ring any bells? Although, whether Bitcoin is a commodity or currency still remains up to speculation. I must mention that these are extremely risky trades and are not for beginners.

Risk Vs. Reward: The Yin and Yang of Investing

Investing is no different from the journey of a student training to become a kung fu master. You see, investing isn't a game of luck or gambling; it's about strategy and calculated decisions. Just as the kung fu master learns to harmonize their body's opposing energies, you must learn to balance risk and reward in your investments.

Each investment vehicle, be it stocks, bonds, or Bitcoin, come with different levels of risk and reward. Risk tells you how much you could lose, while reward tells you how much you could earn. The goal of investing is to make money, right? Therefore, avoiding risk is crucial to reaping rewards. But escaping risk entirely is a fairy tale. No single investment vehicle completely eliminates risk–keep that in mind at all times!

That's why I consider risk and reward the finance equivalent of yin and yang. In the finance world, this balance is measured with a metric called the "risk/reward" ratio. The risk/reward ratio informs you of how much you stand to lose versus how much you stand to gain.

Risk Management: How to Minimize Financial Risk

Budgeting and money management serve as your martial arts training. This prepares you to defend your financial well-being against life's uncertainties. How do you earn your black belt at managing risk? Here are a few drills:

- **Create a risk-proof budget**: Using your budget as a guide, create a risk-proof budget that stirs you away from risky spending habits. Remember, it only works if you stick to it.

- **Have a strong financial foundation**: Financial foundation refers to your mindset and other practices that enforce financial security. A solid financial foundation is akin to building a house with concrete—it is fortified against most natural disasters. This leaves you prepared for any unforeseen challenges that may come. And, believe me, they'll come.

- **Get an accountability partner:** Who'd have thought that having someone hold you accountable for your actions will improve your chances of becoming more financially responsible. Surprisingly, having someone remind you of the mess you made increases your chance of committing to your goals to 95% from 65% (Liska, 2017).

- **Educate yourself:** Keep the fire burning! The world evolves every single minute, so you need to evolve with it or get left behind. There's a lot to learn in the world of finance; for example, you could learn about analyzing stocks. Gather knowledge and use it to fuel your race to success!

- **Diversify your investments:** One golden rule of investment and in life in general, is to never "place all your eggs in one basket." What makes a smoothie so delicious and refreshing? It's the combo of different fruits blended into one! Just like a smoothie needs a blend of different ingredients, you need to keep your wingspan as wide as possible in order to maximize your chances of success.

HOW TO GET STARTED WITH INVESTING AS A TEEN

Warren Buffet is famously known for being the wealthiest investor of our time, but less so for making his first investment at the age of 11. That's right! At age 11, most of us thought only of cookies and chocolate pie, but not Warren. At age 11, He bought six shares of a company called "Cities Services Preferred" at $114.75 dollars. If young Warren could do it, so can you! But how did he do it?

First, you want to understand the basics of investing and ask yourself the following questions:

Do I want to Invest Myself or Let a Professional do This For Me?

If you're investing yourself, then there's some work to be done! You need to invest time to research your choice of investment. After that, you'll need to come up with a strategy; that is, figure out the right time to invest, how much to invest, and in what to invest. Do you have the time it takes to invest in financial education? It's no walk in the park!

On the flip side of the coin, you could hand your money to an expert who'd do all the work for you, but at a price. You don't need to research or develop a strategy. After that is sorted out, we'll move to the next question.

Launching Your First Portfolio

Just like groceries are bought and sold in a market, assets are also bought and sold in special markets. For example, stocks are sold on the stock market. So, if you want to invest in your favorite streaming service, say Netflix, then you head to the stock market and search for the company.

The investment market is organized by companies that sell or buy assets from you. They are called *brokerage firms* or *brokers*. The *brokers* provide the marketplace for you to buy assets from or sell to organizations and other investors.

Where do you find such brokers? Thanks to technology, you can invest with your iPhone or Android device sitting right in your pocket! You can find brokers online in the same way you use Facebook. All you have to do is visit their website and sign up.

Proceed with caution though, and always verify the credibility of the broker before doing business with them. Anyway, I've compiled a list of some trusted brokers who offer packages suited for teens like yourself. They are:

- Fidelity Youth Account by Fidelity Investment
- Greenlight app
- Acorns Early app
- EarlyBird app
- UNest app
- Webull app
- Step app

To invest with all of the above, you need to open an investment account with them and provide some form of government-issued identification, like a driver's license or social security number. If you don't have any of these, don't worry; your parents could assist you by opening a supervised account on your behalf.

BUILDING WEALTH THROUGH COMPOUND INTEREST

Compound interest is one of the most important parts of investment. It's like baking cookies; you start with a batch of dough, which is used to produce lots of cookies. But that's just regular interest. Compound interest is like adding chocolate chips to that dough. That's more sweetness for you!

In the same way your capital earns you more money, interest, which is then added to your initial capital, brings you more money!

Capital = interest = more capital = more interest = more capital...

This can continue for as long as you want.

Let's take a practical example of how compound interest works. Buckle up for some finance technicalities.

Assume you've set up your investment account already and have deposited $600 dollars into your account. After careful study, you've decided to invest in your favorite fast-food eatery, McDonald's. Now, the question is, will this investment make you rich? Let's find out! For the sake of our example, we'll assume McDonald's offers 10% per year interest on your investment.

This means McDonald's will pay you $60 (10% of $600) every year, $120 after two years, $180 after three years, and so on. The really cool part is you don't have to wear an apron and work at McDonald's!

But that's just interest. Take a look at what happens when compound interest comes in:

Comparing Regular Interest with Compound Interest			
Year	Initial capital ($)	Balance after interest only ($)	Balance after compound interest ($)
2023	600	660	660
2024	600	720	726
2025	600	780	798.60
2026	600	840	878.46
2027	600	900	966.31
2028	600	960	1,062.94
2029	600	1,020	1,169.23
2030	600	1,080	1,286.15
2031	600	1,140	1,414.77
2032	600	1,200	1,556.25

Do you see the difference between the "interest only" column and the "compound interest" column? With "interest only" you earn $1,200, while if you allow your money to compound, you earn $1,556.25. That's an extra $356.25!

To find the compound interest, I could also use the compound interest formula:

$P = C (1 + r/n)^{\wedge}(nt)$

'C' represents your initial deposit; that is, $600.

'r' refers to the interest rate, which was 10% in our example.

'n' is when the interest is paid, which is per annum.

't' is the length of the investment, which is 10 years.

'P' is the final amount.

So, using this formula for our example, we have,

P= $600 (1+ 1 year/10%) ^ (1*10)

P = $1,556.25.

You could also do these calculations easily online. Many online tools will plot easy-to-read charts and work out the calculations for you. Check out *www.nerdwallet.com/calculator/compound-interest-calculator.*

The benefit of compound interest is as glaring as the sunshine! Compound interest gives your investment a growth boost, just like fertilizer does to plants. Hold on though; there are other things you need to know.

Time is Money!

From the table above, you must have noticed that it took time for your investment to grow, right? Yes! Your investment grew with time—in the span of 10 years, your capital grew from $600 to $1,556.25. Remember short-term and long-term goals? The magic of compounding interest works best with medium-term to long-term goals. The longer you allow your interest to grow, the greater your account balance. It's that simple!

Let's put everything together now. The secret potion that makes money grow has three main ingredients; interest rate, initial amount or capital, and time. These are the key ingredients you should look out for when planning your investment.

Invest Today and Never Wait Till Tomorrow

Since time is money, it makes sense that the earlier you start investing, the better. There are a ton of reasons why investing sooner than later is a much better option:

- If things go south, you have time to dust yourself and get back up. Losses in the market are common, and in such times, you need time to bounce back strong!
- More time equals more money! We just talked about the value of money and the potential to grow over time—if invested.
- You have lots of room for risks. You tell me, who shouldn't be taking risks between a 50-year-old and

an 18-year-old? Your youth is the time to learn from your mistakes and your risks and grow.

- Starting now gives you more time to save money for the future. You're in a better position to make plans for retirement, education of your children, business goals, and so on.

ACTIVITY 1

If you invest $5,000 in the year 2025, at an annual interest rate of 7% compounded annually, how much will your investment be worth after 5 years? Use the online calculator from *www.nerd wallet.com/calculator/compound-interest-calculator*.

1. In the initial deposit box, input $5,000.
2. Leave the "Contribution amount" box as zero since monthly contributions will be made after the initial deposit.
3. In the "Years of growth" box, insert the number of years for your investment, which is five.
4. Fill the "Estimated rate of return" box with 7% (your interest rate).
5. Since interest is added annually, select "annually" from the "Compound frequency" drop down menu.
6. Finally, click on "Calculate."

The total balance should be calculated as $7,013. This means that your $5,000 investment will be worth $7,012 after five years with an annual interest rate of 7%. There you go; your first compound interest calculation!

END OF SESSION

Gen Z is stepping up its financial awareness game, recognizing the financial challenges faced by previous generations and actively seeking financial education. The great thing is, you're a part of it! The journey to becoming a financially savvy teen starts with grasping the fundamentals, especially when it comes to your mindset.

In this financial world, both saving and investing are vital components for achieving financial growth. Saving involves setting aside a portion of your earnings, while investing is about putting that money to work in wealth-building assets like stocks and bonds. These are skills that require learning, practice, and mastery to yield fruitful results.

However, investing is like a powerful machine, but it's not without its risks. Blindly pouring money into an asset without proper risk assessment is a gamble no one should take. Life itself is filled with risks, but the key is to aim for manageable ones. Investing can be even more potent when combined with the concept of compound interest.

This financial tool has the potential to transform wealth-building journeys. The catch with compound interest is that it thrives on time, so starting early is essential. Doing so offers several advantages, including the ability to bounce back from setbacks, ample time for strategic planning, and more.

Have you wondered why people find it so hard to save money? Flip to the next chapter, where I'll show you the secrets to managing your finances like a finance superhero!

3

NAVIGATING CREDIT AND DEBT

S ometimes, I can't help but feel sorry for the millions of people who are neck-deep in debt. I was reading an article the other day on the level of debt the average American has for the year 2023. The more I read, the sadder I felt. Can you imagine that 35% of Americans spend most of their lives in debt (Cagnassola, 2022)?

Although it was shocking, I was hardly surprised. The main source of debt isn't student loans; it's credit card debt. A whopping 28% of Americans carry credit card balances. What's even more concerning is that credit card debt is just one part of the problem. Americans also struggle with car loans, medical bills, student loans, and more. Debt here, debt there, debt everywhere! And the worst part? Debt is on the rise!

Come to think of it, it makes sense that people owe more to credit card debt. The access to credit cards is right in their pockets—literally! Wait a minute—Do you really know what a credit card is? In this chapter, I'll introduce you to the world of credit cards, explaining how they work and how to avoid falling into the debt trap. Are you ready for the ride? Hop in!

CRACKING THE CREDIT CODE

Who would have thought that a simple piece of plastic could turn into a nightmare? A credit card is a plastic card issued by banks or financial institutions. It allows you to borrow money to pay for goods and services, which, if not paid in full each month, will require you to pay back with interest. Every individual who has a credit card has something called a "credit score."

A credit score is more or less like the grades you get for math class; the difference is, instead of an "A" or an "F," it's a three-digit number ranging from 300 to 850 (The Investopedia Team, 2023). Let's break it down (Equifax, n.d.):

- Excellent credit: 800-850
- Very good credit: 740-799
- Good credit: 670-739
- Fair credit: 580-669
- Poor credit: 300-579

The 300 to 850 score was created by the Fair Isaac Corporation, also known as FICO. It works like this: If you pay your loans on time, your score goes up; if not, it goes down. The credit score helps the banks decide who they should lend money to. If you have a low score, banks may not give you more loans, or they'd do it with a grumpy face.

The credit card could become your best ally or a lurking nightmare that haunts your dreams. It all depends on your FICO score. So, how do the banks know if you've been good with your loans? They calculate your FICO score. The FICO credit is calculated using five factors:

- **Payment history**: This shows if you're paying your bills on time. If you've had late payments, it can lower your score.
- **New credit**: Opening lots of new accounts quickly can make you look risky to lenders.
- **Amount**: This looks at how much of your credit limit you're using. Using too much can hurt your score.
- **Credit mix:** Lenders like to see you can handle different types of credit, like loans, car loans, credit cards, and other forms of debt.
- **Length of credit history**: Having a longer history is better. It shows you've been responsible with credit for a while.

Just like your school grades are reported on a report sheet, all information about your credit score is presented in what's called a *credit report*. The credit report is what financial institutions use to determine whether they trust you enough to give you a loan or not. So, who keeps this credit report? Well, there are three U.S. agencies that handle the report: Equifax, Experian, and TransUnion.

Each of them have their own ways of grading, but they all follow a formula: Payment history takes 35% of the score, amounts owed 30%, length of credit history 15%, and types of credit history 15% (Kagan, 2023). You want to make sure you score high points for each category, and as you can see, payment history carries a lot of weight.

How to Check your Credit Report

The chances are you don't have a credit report yet. If not, there's actually no way to check something you don't have, is there? Or do you have one yet? If so, head on to the *annualcreditreport.com* website to see your performance.

BUILDING A POSITIVE CREDIT HISTORY FROM A YOUNG AGE

Even though you may not have a credit history yet, it certainly is helpful to know how to build a good credit score before you enter the finance world. It makes sense to begin operation "build my credit card score." Doing so saves you the headache of doing it later in adulthood—your attention will be on a lot of things by this time.

So, the question I've been waiting to answer: "How do I build an awesome report?" Well, my friend, follow these steps:

1. Hitch a Ride on Someone Else's Credit Card!

If you can't get one for yourself, get authorized access to the credit card of your parents or a guardian. It'll be better if they also have a good credit history too. Discuss with them to work out getting you access to their credit cards.

This allows you to make purchases in your name and can certainly help you build a credit history. How does it work? When you become an authorized user, you get your very own credit card—that piece of plastic you slot into the ATM. Yes, that one. It will have your name on it too! Hold on though, since the account isn't yours, the account holder will be making the payments, which you can contribute to of course.

2. Get Your Own Credit Card. Duh!

If you are of legal age, you can apply for your own credit card. This comes with a little twist though: Because it's your very first, you get what's called a "secured credit card." To get a secured credit card you have to make an initial deposit of about $200 to $500. Allow me to elaborate.

Suppose you apply for a secure credit card and the lender asks you to make a downpayment of $400, your account becomes activated with an available loan amount of $400. Think of it as a trial period—a test to see how well you'd perform on your first loan. Remember, paying on time is a critical component of your credit report. Pay on time to boost your credit score. Good luck!

3. Try a Credit-Builder Loan

Okay, this is a weird one. Some lenders offer a special loan package called the "credit-builder loan." The purpose of this loan is to help you build credit. Obviously. The weird thing about this loan is that you only get the loan plus interest after you've repaid the loan.

For instance, you take a credit-builder loan of $500 from a bank and repay $520 ($500 plus interest), but the money will not be given to you immediately. You'd have to make monthly installments within a time frame. When you've completed the payments, you then get the $500. Weird, right?

Aside from the steps mentioned above, paying your bills on time is crucial to building a good report. Seriously, paying after the deadline still affects your score. Plus, you may be slapped with a late fee.

The more you meet deadlines, the more benefits you enjoy on your credit card. Not only does your credit score shoot through the roof, but you'll also get more and larger loans at lower interest rates. You'll wander about stress-free knowing your debts are paid and your financial report card is looking dope!

Be Cool, Stay Responsible, and Dodge Debt Traps

In the bustling city of Financia, there lived a young woman named Lily. Having received her very first credit card, she was eager to embrace the world of financial freedom and convenience. But things quickly went downhill from that day.

Lily's first misstep? She opened her mailbox one evening and noticed it flooded with offers from different banks telling her of the amazing perks of their credit card. Lily couldn't resist the offer: "It'll be nice to have more," she said. She applied for some of them, and within days, shiny new cards kept pouring in.

It wasn't too long before Lily realized managing multiple credit cards became overwhelming, leading to the inevitable—missed payments and late fees. She tried paying small amounts per month to cut down her debt, but it only led to more debt. The interest rates were working against her.

She soon realized that she needed help. So, she reached out to her dad's friend, who happens to be a banker. Together, they went through her bank statements, credit report, and her expenses. The banker went, "Oops, habits like this will only get you stuck in the debt trap." Lily asked how she could get out of the pit she had fallen into.

"So, the first misstep was spending on so much fancy stuff. Jesus! Who are you, Taylor Swift?" said the banker. "The number rule of getting a credit card is don't apply for too many at once. For your age, you shouldn't even have more than two."

Lily, feeling reckless, began to excuse her behavior, "I didn't know what I was doing. I never checked my expense statement; I didn't even know I could. I admit, I shouldn't have gotten too many cards. But the bank told me I'd be okay."

"No, Lily, it's not really about having too many cards. That's just one problem; some of these cards aren't the best for teenagers. Even so, I see you got yourself some pretty nice gadgets for your mobile phone. Plus, you've borrowed more than you should on two of the cards already. I know the ads were pretty convincing, but you need to be responsible too."

Fixing The Credit Card Debt Mess

Lily was so deep in the debt trap, her dad needed to bail her out. The financial advisor had taught her the tricks to managing her credit cards:

1. Watch your spending and pick a card that fits your spending pattern:

You don't want to spend without proper planning, right? Remember the lessons from Chapter 1? Don't get a card until you've figured out how your expenses will affect how you use your credit card. Credit cards can vary, with some having lower interest rates, offering more rewards, or providing a higher credit limit.

However, the credit card should serve one purpose—as an emergency plan. Avoid using them for routine monthly expenses like groceries, digital payments, and online shopping, unless it's a genuine emergency.

2. Consider what the credit card has to offer against the cost:

Owning a credit card comes with expenses; annual fees, interest rates, late fees—just to name a few. You want to factor all of these in when choosing the credit card against the benefits; for example, what is the grace period beyond the due date? What reward programs does the credit card offer for meeting payments? Does the card provide advantages for a certain expense?

3. Using multiple credit cards:

Having multiple cards is akin to having different shades of makeup or types of weapons in an arcade game. Not only do you need to spend each type on the right expense, but you should also know how to combine them to work for you. For instance, card A could be suitable for everyday shopping, while card B may best be used for railway transport (Credit cards are your emergency plan).

You can further use the combo of cards to better plan out your finances. For example, paying off credit cards with higher interest rates first will keep you out of the debt trap. The periods in which credit cards offer benefits may also vary. For instance, if card A offers low interest rates on the first week of the month, but card B offers the same but on the second week of the month, then use only card A for week 1 and card B for week 2.

BORROWING IS OKAY—IF YOU'RE DOING IT RIGHT

I understand that no one wants to be in the position where they have to borrow to live, but that's just a stigma. Borrowing plays a vital role in managing personal finances. If you think about it, even some of the world's wealthiest individuals occasionally borrow money when it makes financial sense. Let me chip this in—there's good borrowing and there's bad borrowing. In instances when you need financial support beyond your savings, consider taking a loan—a good loan—as an alternative.

When you take a loan from a bank or financial institution, you'll be required to pay back the borrowed amount with interest. You can choose to repay it in a lump sum or through multiple installments, depending on the loan terms. Now, let's explore the types of loans that might be available to you at your age.

Financial institutions, such as banks, provide a variety of loan options. If you want a loan to buy a computer, cover medical expenses, or handle an emergency, you typically apply for what's known as a *personal loan*. You could use a personal loan to purchase a car, but there's a special type of loan specifically for buying a car. It's called a *car loan*. What loan do you think we'd need for education? The student loan!

DRAWING A MAP OUT OF DEBT MAZE

Wear your dancing shoes because the party is about to get started! Debt is a deep maze; most people get lost the harder they try to get out. That's because they don't have a plan of action–no map to get them out of the maze. If you follow this guide, you'll be able to navigate your way out of the debt trap.

Step 1: Make a detailed list of all the debts you owe–include the dates of collection, total amount owed, and repayment dates. Even though you will not be paying all debts in the same month, it's better to have the overall picture.

Next, work out the percentage of each loan by dividing the amount owed per creditor by the sum of all debts owed. For example, suppose you have a total debt of $26,260, which includes a car loan of $6,000. The percentage of car loan to the total debt is 22.85% (6,000/26,260). I find percentages reflect the magnitude of debt much better.

All Debts For 2023					
Debt Type	Date received	Repayment duration	Amount ($)	Percentage (%)	Interest rate (%)
Car loan	Jan 2023	2 years	6,000	22.85	12.5
Student loan	Jan 2022	6 years	20,000	76.14	5
Personal loan from James for groceries	Jan 2023	2 weeks	60	0.23	0
Credit card	Jan 2023	4 weeks	200	0.76	15%
Total			26,260.00		

Some debts last longer than others and require installment payments. Suppose your installment payment for the car loan is $250 per month, then for the month of January, you owe just $250, and not $6,000. Other loans, like "personal loan from James for groceries," that are short-term will be paid in the same month or as agreed upon. You'd continue to do the same for each month—recording how much is paid, amount left, or you had to postpone. Make sure it shows how much of the long-term debt you've cleared and how much you have left.

Table Showing Debts For the Month of January 2023						
Debt Type	Date received	Total amount owed on loan	Amount due monthly ($)	Total outstanding debt (%)	Interest (%)	Outstanding amount owed ($)
Car loan	Jan 2023	$200	$20	30.30		
Student loan	Jan 2022	$300	$70	45.45		
Personal loan from James for groceries	Jan 2023	$60	$60	9.09		
Credit card	July 2023	$100	$10	15.15		
Total per month		$660				

From the table above, our monthly debt is $660. Isn't that less frightening than $26,260?

Step 2: Work out your plan of attack. Don't forget, different types of loans come with different interest rates or penalties. If your income can pay for all loans without affecting your lifestyle, then go for it. However, what if you don't have enough funds? What is the line of action? There are several ways to take this head on:

1. The avalanche debt payoff method

An avalanche begins from the top and crumbles to the bottom of the mountain. The mountain represents your total debt, and the snow represents the individual debts you owe. The avalanche method is a strategic way of tackling your debts. And just like an avalanche, you crush your debts from top to bottom. However, you're not just picking the debt with the largest amount, you're also picking on the high-interest debts first.

Based on the "All Debts For 2023" table above, which debt would you clear first using the avalanche method? The credit card, right? Awesome! The advantage of the avalanche method is that it saves you the most money on interest payments because you're tackling the most expensive debt first. However, it may not necessarily tackle the highest amount (credit card loan is smaller compared to the car loan).

2. Snowball debt payoff method

This is actually my preferred method. The snowball method does the exact opposite of the avalanche method—you start from the bottom and work your way to the top. In addition, you won't be considering the interest rates with this method.

From a psychological standpoint, it's easier to settle smaller debts than larger ones, right? You'd have more money to spend! The excitement from quick wins further motivates you to tackle more debts! The downside of the snowball method is that you end up paying more interest rates.

So, now, from the "All Debts for 2023" table, which debt has the lowest amount? "Personal loan from James for groceries?" Exactly!

3. Earn more money and cut down debt

The faster those debts get settled, the better. In addition to the debt payoff methods above, you could adopt healthy money habits that'll turbo-charge you out of the debt trap. There were certain times when I felt like I was stuck in quicksand. The more I struggled to get out, the faster I sank. Burying my head in a pillow and screaming my lungs out only improved vocal cords. My debts didn't go away. It was at this time that I asked myself, "What do I need to do to escape this pit?"

One thing was obvious; my debt payment plan wasn't working out. Why? I needed more money, obviously. How does one get more money? Two options: One, I earn more money; two, I spend less. At the moment, option two was the only feasible option. Spend less it was. Although I was in the mud, my new plan was coming to my rescue. "I can do this," I thought.

4. Bury your credit card

While you cut down on expenses, be sure to not stack up any more debt. Avoid the temptation to use the credit card; remove it from your wallet if you must! Clear all your credit card information from online stores.

5. Look into debt consolidation

Debt consolidation is another strategy to look into when others fail. For instance, suppose you have a car loan, education loan, and others. As you know, each debt comes with its own interest rates and due dates. That's a war on too many battlefronts. What if you could combine all these loans into one? That's where debt consolidation comes in. Debt consolidation is a loan, called the "debt consolidation loan," given by a financial institution that can be used to pay all other debts.

Now, you have one battlefront to fight. Instead of making payments for different loans, you now have only one loan with one monthly payment and even affordable interest rates.

Don't Make the Same Mistake Twice!

Cleared your debts? Don't relax just yet. We've only defeated one enemy. The next challenge is mustering enough control over your finances to not end up back where you started. Some healthy finance habits to remember are:

- Only purchase things you can afford without using a credit card. If you can't afford it now, it's probably not the right time to buy it.
- Save money for unexpected emergencies, like a sudden car repair or medical bill. It's your safety net.
- Do your best to pay off the entire credit card bill every month.
- From your budget, mark out things you need versus what you want. Focus on covering your needs first.
- Create a budget to track your spending and save the things you want.
- When you get more money, like a raise or gift, think about saving or investing it instead of spending it all.

It might not seem like it, but giving yourself a pat on the back for goals accomplished is an effective way to stay motivated.

END OF SESSION

I'd prefer not to talk about debt at all; I would love to avoid it at all costs. But life isn't a Walt Disney cartoon where every tale ends with a happy ending. Nonetheless, there's an age-old adage that says, 'Cash is King,' and indeed, cash reigns supreme. Do as much as you can to avoid going into debt. I recommend tuning in to "The Dave Ramsey Show" podcast or grabbing his book *The Total Money Makeover* for invaluable lessons on staying clear of debt. Good luck!

Falling into debt is very easy, but getting out needs careful planning and dedication. With the right methods, you can overcome it and enjoy the joy of financial freedom. One very easy trap to fall into is the credit card debt trap, which has a lot of people in its clutches. That said, it's worth noting that credit cards in particular wield substantial financial influence when used wisely.

Your credit score tells the world how good you are at managing money. It ranges from scores of 300 to 850 and is influenced by factors like payment history, new credit, amounts owed, credit mix, and length of credit history. Without a doubt, it's a smart move to build a good credit score as early as possible.

When your savings can't meet up with your expenses, borrowing is another viable option. But it's important to be smart about loans and take care to understand the rules. Debts often grow quickly, and in such cases, using either the avalanche or snowball method can be effective in managing multiple debts.

Remember, with the right mindset and strategies, you can avoid the chains of debt and secure a brighter financial future.

Now that you've mastered the basics of credit and debt, let's dive into exciting opportunities to earn money, both through part-time jobs and entrepreneurship. Ready to start making money work for you?

4

EARNING MONEY—PART-TIME JOBS AND ENTREPRENEURSHIP

Gen Z may just be the most fascinating bunch of us all: There hasn't been any generation so determined to change their lives for the better. I realized this when I came across a survey by Junior Achievement USA—an organization dedicated to inspiring young achievers like you—I was blown away. They reported that a staggering 60% of today's teenagers have a burning desire to grow their wealth (Reinicke, 2022). That's just the tip of the iceberg! Gen Z are rolling up their sleeves and plunging into the corporate world—they aren't waiting for adulthood to make it happen.

Grab your highlighting pen because the wisdom you'll gain here can change your life's trajectory. In the following few pages, our mission is to transform you from a dreamer into a force to be reckoned with within the world of work and business. We'll unlock secrets of creating a resume that shines, crafting a cover letter that demands attention, and diving into the exhilarating world of entrepreneurship. Are you ready? Let's go!

POLISHING YOUR CAREER WITH A RESUME AND COVER LETTER

You only cross the bridge to adulthood when you're able to land your first job. But you can't waltz into an employer's office demanding a job. No! You have to prove to the employer that you're at least capable of performing your duties. "How does that happen when I've not gotten a job yet?" Good question. That brings us to a document called a "resume." A resume is a document (one to two pages) that informs the employer that you're responsible and capable of getting the job done.

Employers carefully analyze the resume to determine if you're the best fit for the position. Hence, you need to make sure it presents your capabilities flawlessly. So, what is a resume made of? A basic resume contains the following essential sections:

- **Contact information**

Begin your resume with your contact information; your full name, mobile phone number, email address, residential address, and even a social media account like a LinkedIn profile.

- **Profile summary and objective**

This usually forms the header or otherwise goes at the top of the resume. It's usually a short paragraph that tells employers what you're aiming for and why they should be interested in you. It's like your personal sales pitch. For instance, say you're on the lookout for a summer job at a nearby restaurant. Here's how your summary may look like:

> "A hard-working high school senior with two years of experience serving in the school cafeteria. I am passionate about great food and exceptional customer service and eager to kickstart my career in the restaurant industry. I'm excited about the opportunity to join a dynamic restaurant team."

Before you write the profile summary, always study the job description: It'll provide insight into what the employer expects. What skills and relevant experience are relevant to the position? Is the job a fit for you? Do you have the right qualifications? And so on.

Use the first sentence to introduce yourself as a suitable candidate for the job. Mention your current status with some personal traits that align with the requirements of the job. Then, proceed to inform the employer of your qualifications, which are relevant to the job. Use catchy keywords that the employer used in the job ad. As in the example above, notice keywords and phrases like "hard-working," "passion for great food," and "exceptional customer service."

- **Education**

This section tells the employer about the qualifications and educational background you have. Again, this should be relevant to the job you're applying for. This should include the name of the school or organization, the year of completion or graduation, and the qualification itself. In addition to your high school diploma certificate, try to get certifications that align with your career choice. There are online platforms that offer tons of courses; for example, Coursera, Udemy, edX, and so on. Also, look up apprenticeship opportunities. The more qualified you are for the job, the greater your chances of securing the job. For example:

Education and Professional Qualifications	
• High school Diploma (GPA: 3.68)	2022
Name of school or organization	
• Diploma in Catering	
University of Michigan (via Coursera)	2023

- **Experience**

Have you worked in a similar setting as the job you're currently applying for? The experience section lets the employer know that you have performed this job before. If you've served as a customer service representative in a previous job, performed a similar role in high school, helped a family with a similar job, or gained experience from volunteering and freelancing services, then it goes here. Since this is an entry-level job, the employer might not be too critical about past experience, but someone out there might have more than you. Most times, the job goes to them. That's why you need to stand out. You'd present the organization where your experience was acquired, the time frame, and the activities done.

<u>Experience</u>
 Gryffindor college
 (2020–2022)
- Before lunch time, I cleaned and decorated the school cafeteria environment.
- Assisted in dishing out and serving the meals to students from the buffet pans.
- Tidied up the tables—sent used plates and utensils to the kitchen after lunch period.
- Helped students choose the best combination of cereals, milk, proteins, and vegetables to take.
- Relayed student complaints to the head cook and communicated back to the students.
Extracurricular activities
- Assisted in parceling and serving food and drinks at community cookouts.

Like your objective summary, you want to use job-related keywords. Can you spot the catering services keywords in the sample above? Finally, ensure all experiences listed are related to the job you're applying for. Do this and employers are sure to take notice of you.

- **Skills**

This section lists the skillsets you have that will allow you to perform the job. Skillsets are categorized into two; soft skills and hard skills. Soft skills are those that come naturally to everyone, although they can be worked upon. For example, teamwork, time management, and so on. While hard skills are those obtained from hands-on experience or education. For example, writing, reading, Microsoft Word, and so on.

- **Interest and hobbies**

Now comes the fun part. The interest and hobbies section show the employer what you do in your free time and whether they are things that further impact your skillset. For instance, a fiction-writer who loves to read fiction will be a better candidate than a fiction-writer who loves to swim. It tells the employer you love what you do.

- **References (Optional)**

The references section is optional; however, employers may request that you provide the contact(s) of a referee who can vouch for the claims you've made on your resume. If you say you've worked at company A, the employer may want to verify that claim. If you prefer to include a reference, you provide the full name, job title, company or organization, and contact of the individual. Like so:

Reference
John Doe
Professor
University of California
555 555 555
Johndoe@gmail.com

The Cover Letter: Show Them What You're Capable of

The resume letter provides an overview of your ability and is used to draw the attention of the employer. Okay, so, you've got the employer's attention. Now, it's time to convince them to choose you! That's where the cover letter comes in. The cover letter further highlights your skills, qualifications, and experience. You're not just listing your skills as in the resume, but you're also trying to convince the employer that you're qualified for the job by using your experience as your selling point.

Writing a cover letter begins with carefully reviewing the job ad itself. Gather the keywords used in the job posting and ask yourself if you're qualified enough. You should be able to tell the employer that you have great command of the skills requested for or are at least able to catch up as soon as possible.

Begin with a salutation such as, *"Dear Human Resource Manager," "Dear Sir/Mam,"* or *"To Whom It May Concern."* Then move to the next line. Introduce yourself, ensuring you express your excitement for the job opportunity.

"My name is Jane Doe. I'm a recent high school graduate from Gryffindor College. I'm seeking an entry-level role as a customer service representative at your organization. I'm confident you'll find my skills and experience most beneficial to your company."

After the introduction, blast the employer with feats of accomplishments and experiences!

"I was a very active member of the High School Live Podcast and worked as a presenter for two years. I specifically handled sessions that addressed complaints from students about school activities. I took phone calls from students and teachers alike and mediated conversation between both parties to ensure a resolution was reached.

Also, I have practical knowledge about setting up a studio to achieve my objectives. I often assembled the production equipment such as the microphone, video cameras, and lighting for the production. Also, I used software such as Audacity, Windows Video Editor to edit audio and video footage from podcasts which I then upload to several streaming platforms."

After a grand display of your capabilities, you want to close the cover letter by asking for an opportunity to meet the employer for an interview.

"Thank you for your time and consideration. If you need to further discuss how my skills and experience will serve your company, I'd be glad to attend an interview at your appropriate time."

You want to then close the document with your signature and name at the bottom right or left-hand side of the document.

"Signed,

Jane Doe"

INTERVIEW PREP

Finished crafting your resume and cover letter? Don't relax yet; there's more work to be done. One summer afternoon, I stepped into a charming coffee shop during my lunch break. I couldn't help but eavesdrop on the conversation unfolding at the table behind me. Two people were talking, and their voices were hushed but tinged with concern. They were discussing an unsuccessful job interview.

"Mark. So, you remember that guy we interviewed last week, right?" the taller of the men asked.

"Yeah, the one for the open marketing position? What happened to him?"

"Well, we couldn't hire that guy, though," said the taller man.

"Oh, really? What went wrong?"

"The boss felt he lacked communication, or rather, couldn't communicate his ideas. Remember, he just kept stuttering?"

"Yeah! I remember. Too bad; he seemed like a nice guy and had the qualifications of a fine marketer. Well, for a job like ours, communication is key, after all. So, did you hire someone else instead?"

"You bet! And trust me; although not as qualified as the first, this guy is a huge asset. He came in with a vintage watch and convinced the boss to buy it. He told the most fascinating story to do it! It's a good thing he's on our team."

"Interesting! I'd have to test him myself to believe you!"

"I'm telling you, Mike, we just might have another Jordan Belfort on our team. Ha ha ha!"

There was a brief and awkward silence.

"Oh, come on, don't tell me you've forgotten the movie *The Wolf of Wall Street*?

"Ha ha ha! I get it now!"

My time was up, and I left the coffee shop, but their conversation lingered in my thoughts. It was a subtle, yet powerful, reminder of the importance of effective communication and, of course, best practices in securing a job.

The first candidate they interviewed had the better qualifications but failed to express those qualifications verbally. Even though the less qualified candidate got the job, it's obvious the first candidate was unprepared for the interview. Hence, they lost the opportunity.

The very first tip is to always do some background research about the company. Find out their objectives, their mandate, and so on.

Interviewers also come prepared for the interview. What do I mean? Good question. The interviewer is determined to select the right candidate. Yep. So, they have a list of carefully crafted questions that throw most candidates off balance. For example, the most daring of them all: "Tell me about yourself." What is the interviewer asking? My height, likes and dislikes, my family's background? What is it exactly!

Sorry to burst your bubble, but the question is asking for a story about yourself that's beneficial to the role you're interviewing for. Talk about what made you choose the career path, why you chose to apply to their company, your passion for the job, your experiences, and how they all make you the best fit. Amazing, right? Yep! That's the golden secret!

One other more straightforward question is, "How did you come to hear about this position?" Be honest here. Note: Never ever lie! If you found them on LinkedIn, say so. But don't just say "LinkedIn," say:

"I've actually been following your company's account on LinkedIn for a while, hoping a position would turn up. I looked up the job description and realized I have the skill set required and the experience to take on the role. So, I was more than excited to apply for the position."

Which one sounds better?

Other questions to be on the lookout for are:

- Can you handle the pressure or cope with stressful situations? You can share a previous experience where you remained calm and calculated during a chaotic situation.
- Are you a team player? Inform them of positions you've had in the past that involved cooperation with

multiple individuals. A school project is a great
addition!

Obviously, we can't tackle all the questions. You have to do
your homework. Some questions are also industry specific, so
you should bear that in mind.

Your Business Persona

Have you seen the way news presenters speak on the news?
Ever hear anyone use that same tone? I haven't. You see, that's
their business persona! As soon as you step into the employer's
office, you're a different you! A you that's ready for the business
world. Building a business persona is adopting a professional
outlook: A way you conduct yourself in a business environment
and with other business associates.

Assume you're in a movie-verse, let's call it "Financewood";
Hollywood, but for finance. As a skilled actor in Financewood,
you ought to be a critical thinker and an avid planner, among
many other traits.

Don't forget the dress code! Even your language needs change.
Which sounds more professional? "Hey, man! I sold the stuff
for 500 bucks. I made lots of money," or "I was able to sell each
item for $500, which yielded a profit of 10%? I could have said
"made profit" but, "yielded" shows I'm more business savvy.
So, who'd you be willing to do business with? The guy in jean
pants and a baggy T-shirt or the man with a perfectly tailored
suit?

Do you see why your Financewood character is very important? Professionalism is your lifeline in the business world. No exaggerations!

NOT FIT FOR THE 9-5? HOW ABOUT A BUSINESS?

In the neighborhood of Edinburgh, Scotland, people opened their doors to find a 14-year-old Doherty with jars of jams made from his grandmother's favorite recipe. Doherty believed grandma's recipe contained a touch of magic, and it seemed all of Scotland thought so too. By 2007, Doherty had his jam products on the shelves of supermarkets across Scotland and in other parts of Europe. Even Queen Elizabeth II recognized Doherty's remarkable success and awarded him a Member of The British Empire medal. To this day, Doherty has sold more than five million jars of jam.

Now, the question arises: Was it merely a delicious jam that led to his success? Doherty could have said, "Oh, I can make jams," and headed to apply for a position at a jam factory. He didn't do that. Instead of the 9-5 path, he chose another—the path of entrepreneurship.

The Path of Entrepreneurship

Remember I said Gen Z are picking up the pace when it comes to changing their financial status? Every day, young individuals like 13-year-old Hart Main are coming up with fresh ideas and improving products and services.

Hart, for example, started his own company, ManCans, selling scented candles for men. His journey began when he needed money for a $1,200 bike to compete in triathlons.

Mart's sister, Camryn, was already in the scented candle business which she ran for a school fundraiser. He often teased his sister about it and complained that the candles were too "girly."

Little did he know it was about to change his life and that of thousands of others. Main had a fresh idea; instead of "girly" scents, why not "manly" scents? The genius idea was to give the candles a more masculine vibe, so, instead of putting his candles in glass jars, he put them in soup cans. Can you believe it? Soup cans!

At this point, Main has shown two important aspects of entrepreneurship; first, he developed a business concept to sell "manly" candles. Next, he gathered the skills to produce scented candles. But it didn't end there. There are other skills Main used in taking his business to the top. Let's take a look at how to develop a business plan.

Creating Your First Business Plan

An entrepreneur sees the world like a chess game: They need a strategy to win. Main didn't just make the scented candles and hoped he'd be rich someday. No, there was a carefully thought-out plan. First, it was the goal: Main's goal was to raise $1200, remember? How was he going to do this? There are questions to answer before you set up your business plan.

1. Where Do I Get Funds?

The first challenge was raising money to fund the business. He solved this by using his entire savings of $100 to fund the production process. Do you see the benefits of savings? Every business needs funding, which you could obtain from friends and family, loans (if you have a good credit score), or people who share the same interest as you and are willing to help fund your project.

2. Who's My Target Audience?

Products and services are produced for people, right? But it can't be everyone. So, you need to define who your business is targeting. People who show interest in your product and services are your customer base. You need to answer if your business targets adults, teens, or children? Could it be gender specific? People within a certain location? Do you recall Main's customer base? Right, boys!

3. How Do I Market My Products?

Everywhere you look, there's one advertisement or another. Those ads are run by businesses trying to reach a wider audience. Remember Doherty who sold jam jars? What was his marketing strategy? He went knocking door-to-door. Lucky for you, there's social media, which makes marketing tools accessible and easy to manage. Facebook, Twitter, Pinterest,

Snapchat, Instagram are the most commonly used amongst entrepreneurs.

4. What Will My Finances Look Like?

Before you launch the business plan, you need to ensure it aligns with your goals. Main's goal was to raise $1,200 only to fund his triathlon. The business setup was going to provide that amount. To keep the business operational, it should provide a steady flow of cash. If not, then it's doomed to fail even before it starts.

Answered the questions? You can now proceed to creating a business plan.

Step 1: Dream Big—Define Your Mission and Values

Imagine where you want your business to go. Do you see it changing the way people do things or solving a big problem? That's your vision. What's the reason your startup exists? How will it make the world a better place? Your mission statement will answer these questions. Finally, what are the values of the startup? What are the things that matter most? What do you stand for; is it innovation or making people's lives easier?

Step 2: Summarize Your Startup's Big Ideas

This is a short summary of your startup that includes the cool stuff you're going to do. Start with a one-sentence introduction that explains what your business does and how it helps people.

What problem are you solving? How are you going to do it? Just imagine you're writing it to a friend—keep it simple.

Step 3: Set Goals and Achievements for Your Startup

Every game has rules and goals, right? Same with your startup. Set some goals—things you want to achieve. Want to make a certain amount of money? Want more customers? These are milestones to reach. Think of them like checkpoints in your startup adventure!

Step 4: Craft a Cool Company Description

Tell everyone who you are! What's the name of your startup? What do you do? Why is it awesome? Write it down. And if you're working with a team, introduce them too. This is your company description, the part where you say, "Hello world, we're here!"

Step 5: Explore Your Startup's Potential Fans

This goes back to knowing who your customers are. Who are they? What do they like? What are their problems, and can you solve them? That's your market analysis; the key to getting fans for your startup company!

Step 6: Team Up—Find Allies and Resources for Your Startup

No superhero fights crime alone, right? Same with startups. You might need help or resources. Who can you team up with? Investors, partners, suppliers—they're your startup allies. Together, you can conquer anything.

Step 7: Plan How to Promote Your Startup

It's showtime! How are you going to tell the world about your startup? Earlier on we talked about defining your customer base, right? Your marketing strategy should involve how you intend to reach this customer base. What's your unique selling proposition? What platform is your customer base on, Facebook, X (formerly Twitter), Instagram, or Snapchat?

Once you're done with that aspect of things, it's time to launch your business. But, before then, you also need to sort out the legal side of things. This also depends on the scale of the business and type of product or services you're offering to your customers.

SCHOOLING AND WORKING ON THE SIDE

There are lots of students who have to work and attend school simultaneously. One such individual was my friend, Emma, who was working part-time to help fund her college education. I'd asked her how she coped with school and work.

"So, Emma," I inquired, "How do you manage school and work so smoothly?"

Emma flashed a confident smile: "It's about finding your groove. The important thing is to make sure school activities are settled. That's the priority."

"I guess so. Must be exhausting, right?"

"Yeah, sometimes. But I planned my days ahead. That way, I know exactly how much of a work load I have for the day." She paused for a bit and continued: "First, I lay out my week on a planner–school, shifts at the coffee shop, study times–all mapped out. When tasks seem daunting, I break 'em down into smaller bits. It's less overwhelming that way."

Another brief pause: "For school prep, I do it the night before. I have my backpack already packed and lunch already prepared. It takes away the stress of morning prep."

"Must be tough. I'm curious; how did you manage time?" I quickly chipped in.

"I use my smartphone for reminders—assignments, shifts, everything gets a digital nudge. Sometimes, I feel over-whelmed. Like–for real! But I can't afford to drop out. Luckily, I run evening shifts only five times a week. On days I don't go to work, I do more schoolwork to lessen the burden."

"I see. I'm guessing you use a calendar app to schedule all your activities?"

"Yes! That's it! I already know my activities for the week. So, I just set up reminders on my phone and get updates."

"That makes sense," I responded, reminiscing on my own experience. "Screen time can be a tricky one. How do you manage it?"

"Urgh! It is the most difficult—I mean, I could barely focus since I studied with my laptop or my phone. Sometimes, I kept my mobile device locked in my school bag, turned off Wi-Fi on my laptop, and worked offline."

"I see. You've planned your activities remarkably!" I said, nodding my head.

"Yep! My motto is that every second has to count for something!"

Working and attending school simultaneously is not so different from living a double life—not much different from the story of Peter Parker from the *Spiderman* franchise. The key is using every tool at your disposal to juggle both worlds. Also, you should consider the type of job. According to studies, dedicating more than 15 hours per week to your part-time job takes a toll on your performance (Understood, n.d.).

UNDERSTANDING BUSINESS FINANCES

Money management is the foundation of success for entrepreneurs. It's not just about making money; it's about how you use it wisely to make the right business decisions. What do you think happens when money is handled badly? The business collapses. When you understand how money works, you're in a better position to make decisions and set yourself up for long-term success.

What are Financial Statements?

If you want to know how to operate the financial side of a business, one crucial record to understand is the financial statement.

A financial statement is like a diary of all the financial activities that take place within a company. It tells you how much you make, how much you owe, and how much you spend to run the business and much more, including:

- income statements
- cash flow statements, and
- balance sheets

Every solid business has all of them. For instance, let's take a sneak-peek at the income statements of Starbucks. Visit *www.macrotrends.net/stocks/charts/SBUX/starbucks/financial-statements*

Income Statement for Starbucks 2021-2022		
Categories	2022	2021
Revenue	$32,250,300,000	$29,060,600,000
Gross Profit	$21,932,900,000	$20,321,900,000
Operating Income	$4,617,801,000	$4,872,100,000
Net Income	$3,281,600,000	$4 4,199,300,000

Fascinating numbers up there! You can see, Starbucks made a revenue of $32.25 billion in 2022 and made a little under $30 billion in 2021. Revenue, gross profit, operating income, and net income are crucial records that all businesses should have. It is from these reports that you know the situation of your business. Okay, let's dive deeper into the parts of a financial statement.

Income Statements

An income statement, also called the "profit and loss statement," is like a report card, but for businesses. Like that of Starbucks, is a record that tells you four things about your business, like:

- how much money your business generates, referred to as "revenue,"
- the cost of running your business, referred to as "expenses,"
- the profit made from the business, referred to as "net profit" or "gains,"
- And, the losses over a specific period of time.

Sounds like a powered-up budget plan, right? Whether you're running a lemonade stand or thinking about your future career, it's important to keep track of how much you're making or losing. That's the main function of the income statement.

Cash Flow Statements

A Cash Flow statement is a record that tracks where your money is going and coming from. Think of it like keeping track of your allowance or pocket money. For example, you get $10 from your mom every week. That's money coming in, right? If you spend $5 on a snack, that's money going out, right? Write those on paper, and you have a cash flow statement. Here's the cash flow statement from Starbucks.

Starbucks Cash Flow Statements		
Records	2022	2021
Net Income/Loss	$3,283,400,000	$4,200,300,000
Cash Flow from operating activities	$4,397,300,000	$5,989,100,000
Cash flow from investing activities	-$2,146,300,000	-$319,500,000
Net Cash Flow	$-3,637,300,000	$2,104,800,000

From above, Starbucks made a net profit of $3,283,400,000 in 2022 and over $4 billion in 2021. This means their profit for 2022 is lesser than it was in 2021, right? Looking at other records;

- Cash flow from operating activities tells you how much money the company has spent to carry out the business or made from the business throughout the period.
- Cash flow from investing activities records how much money the company spent in buying or selling investments. Notice that it's a negative number? That tells us that Starbucks used $146,300,000 for investment purposes.

The cash flow statement records it all!

Balance Sheets

A balance sheet captures your business's financial situation at a certain moment, like the end of a month or year. It's focus on three main parts:

- Assets (All your stuff): This sums up the value of everything you own. It includes your cash, what people owe you, the things you own, and even money you're expecting in the future.
- Liabilities (Debts): This is the money you owe to others. For example, if you borrowed money from your parents to run the business, that record goes here.
- Equity (What you have left): The equity records state everything you have left after you've settled your debts and liabilities.

Let's look at the balance sheets of Starbucks:

Balance Sheets of Starbucks		
Record	2022	2021
Total Assets	$27,978,400,400	$31,392,600,000
Total Liabilities	$36,677,100,000	$36,707,100,000
Shareholder Equity	-$8,698,700,000	$31,392,600,000

So, if you want to know how much you have, how much you owe, and how much you have left if you settle all your debts, the balance sheet is the record you prepare.

Fundamentals for Calculating Costs and Setting Pricing For Your Goods And Services

Before you hit the market, it's crucial that you understand how to calculate costs and set prices of your products or services. You want to run your business and earn profits, right? Then, you should know the fundamental principles that will help you know how much it takes to create your goods or offer your services.

In addition, you have to determine how much you're offering your goods or services to your customers for, such that it covers the cost of running the business. But, not just the cost, it should also provide you with extra—good old profit!

These fundamental principles are:

Cost of operation

Calculate the cost of all processes involved in making the product. You need to consider the cost of all the materials you use. For example, if you're making custom t-shirts, the materials include fabric, ink, and any special additions like buttons or tags. Don't leave anything out.

You're not waving a magic wand for the shirts to make themselves. You are putting in back-breaking work—no matter how little it is—it should be considered when calculating the cost. Remember, your time is money. If you employ someone, then the wages or salary paid to them must be considered.

Lastly, aside from materials, you're using up utilities, equipment, workspaces, and so on to run the business. You'd also have to consider the light bills, water bills, and cost of equipment.

Adding profit

Once you've figured out the cost of running the business, you can then add the profit. In this case, you get to decide how much profit you want to make on each item. It's from here that you have room to offer competitive prices to your customers. Suppose it costs you $1000 to make your product. You can add a profit of 15% or $150; bringing the prices to $1150 ($1000+$150).

Calculating cost by quantity

If you're making or selling more than one item, there's an extra step to take in calculating the cost and profit for each individual item. It's a simple formula:

- Take the total cost of making the product (don't forget utilities, equipment, and so on). Let's assume it costs you $900 to make 150 T-shirts. Divide the total cost by the number of items. This gives the cost made in making one item, which is $6 ($900/150)
- Now, for the profit: Suppose you want a profit of $300. Divide the profit by the number of items. Hence, the profit per T-shirt is $300/150, which results in $2 per T-shirt.

See? That wasn't so difficult, was it? Next, you stamp on the prices of your goods or services.

Setting your price

Now that you know the cost-per-item and the profit you want to make for each one, it's time for the grand finale—setting the price of your goods. The formula is also simple:

- Add the cost-per-item ($6) and the profit-per-item ($2). This gives us our selling price, which is $8 per t-shirt.

Congratulations! We just figured out how to run our company's finances. But we've yet to cover other crucial areas. For our next discussion, we'll explore how to save our company from financial doom.

Managing Profit and Navigating Losses

As you operate your business, you're also keeping records. Can you remember the types of financial records? The financial statement! Yes! You'd have an income statement, a balance sheet, and a cash flow statement, right? Awesome! Together, they give you every detail about all the financial activity of your business.

Putting your t-shirts in the store is just your starting point. After a few months to a year, you want to check up on the sales and performance of the business.

So, how do you know you're profiting or losing? If you're selling all your T-shirts at $8 then, you're swimming in profits. But, if no T-shirts are sold, or the customers want a cheaper bargain, then, you'd be losing money. How do you know though? You check the income statements.

Remember the "net income" of the income statements? It tells you how much profit you've truly made after taking out production and other expenses. But that's not all. You're also going to combine information from your income statement with others like the cash flow statement to see if you're spending too much on production. Are sales coming in? If they are, where's the money going? Are expenses more than money coming in?

If all these numbers are giving you a headache, then, your best bet is to seek guidance from a financial advisor.

Long-term Growth and Expansion Plans

Remember our T-shirt business where we made each shirt for $6 and sold them for $8? Well, it's time to figure out how to take that business from our basement to our very own factory and grow our T-shirt empire!

Focus on What Works

Maybe you discover that a certain brand of your T-shirts, ones with tie-dye, was a big hit among young adults. What do you do next? Focus on producing more awesome tie-dye shirts. You could try different ideas to see what works too. What's important is that you keep an eye on what your customers love the most. What are they saying about your products? Do you need to rebrand?

Technology is Your Ally

Technology is the other thing that makes the world go round. There are many tasks that technology can lift off your shoulders. For example, you could use software to help you track your inventory and even suggest new project ideas. AI is making a lot of rounds on the news for a reason. You know that saying, "Machines make life easier?" That's old school; today, technology makes your life easier. You could build a website with a few clicks and create an online store to sell your products. The power of technology truly knows no bounds.

Build a Strong Team and Network

Even though you might be the creative genius behind your products, you can't do it alone. Entrepreneurship isn't always a one-man venture; you'd need the skills of other people to help improve your business and upscale it. Build a strong team of individuals who share your ideas to help you handle some aspects of the business while you handle the part you love most. For instance, you can't work production and work in finance at the same time; it's exhausting.

Always Solve Problems

Entrepreneurship ends when you no longer experiment with new ideas that solve existing problems. By staying creative and adapting to changing trends, you'll keep your customers on the edge and coming back for more. Do you see what Apple does with its iPhones? Each release comes with solutions to past problems, new designs, and products.

So, there you have it, future business tycoon! You now have the fundamental knowledge of running your very own business. Embrace the change coming and shoot for the stars!

You've taken control of your financial future by learning how to earn money through part-time jobs and entrepreneurship. Now, let's explore smart ways to make your hard-earned money go even further in the next chapter on smart shopping and saving techniques.

Your Financial Superpowers

"The biggest superpower you can have is the ability to change your own life."

— HRITHIK ROSHAN

Maybe you rolled your eyes at my use of the word "superpowers" in the introduction. How is the ability to shop smart a superpower? Well, it might not be anything akin to what Spiderman can do, but it truly is a superpower. All a superpower is an exceptional ability – so you can have one in anything... and the beautiful thing about having superpowers in personal finance is that it opens doors for you in every direction.

When you have a strong foundational knowledge of how to manage and grow your money, you're able to do so much more in life – and without a lot of the stress that most adults face. Have you ever overheard your parents discussing their money worries in hushed voices? That's the kind of stress that this knowledge you're acquiring now will help you avoid. Now, tell me that's not a superpower!

Your school will teach you all kinds of useful stuff that will help you over the course of your life, but sadly, personal finance isn't one of them. I'm writing this book to bridge that gap and help you avoid and overcome the challenges many of the adults in your life faced when they were growing up. But I'm going to need your help in spreading the word... and the good news is, you don't need special powers to do that. All you have to do is write a short review.

By leaving a review of this book on Amazon, you'll help other teenagers find all the training they need to uncover superpowers where once they lacked knowledge.

This information is crucial, yet for some reason, it's not taught, and far too many people lack the understanding they need to make their money work for them. Together, we can change that.

Thank you so much for your support. See? You're already putting your superpowers to good use!

Scan the QR code below to leave your review!

5

SMART SHOPPING AND SAVING TECHNIQUES

E ver found yourself wandering through a store, shopping list in hand, only to walk out with items you never intended to buy? Or perhaps, you were diligently studying, when all of a sudden, you're craving a can of coke? What if I told you that there's a fascinating, yet somewhat sinister, phenomenon at play in our minds? This phenomenon is a big deal today!

Recall that Americans have problems managing their budget? This phenomenon is one factor. In fact, Americans spend an average of $276 every month on stuff they didn't intend to buy. If you do the math, that adds up to an extra $3,312 spent every year—that's on stuff we didn't really need (Cruze, 2021). Fascinating, yet scary!

What could have the power to compel humanity to do its bidding even to the extent of running into debt? One word:

Advertisements! Have you also noticed that the ads you see on your social media feed are things you're interested in? Ponder that.

You see, people tend to buy what they are familiar with. Promoters often say, "A visible product is the product that sells." Based on that, advertisers ensure that their products are "sparkly" enough to gain attention. For instance, candies have colorful wraps, cars have sleek designs and attractive colors, and luxurious clothes are worn by influencers and advertised by gorgeous looking models.

Advertisements are made to trigger your emotions, which in turn, causes you to give out your hard-earned money. But you want to be stingier with your money, right? What then is the best way to not give in to advertisements? Shut them off? Impossible! Avoid them? Impossible! The burden rests on you to resist seductive advertisements.

WHICH WILL IT BE: THE RED PILL OR THE BLUE PILL?

Do you want to uncover the tactics companies use to entice you and break free from compulsive spending? I present two options: The blue pill, where you continue to spend without restraint and risk falling into debt's grip, or the red pill, where you awaken to economic reality—a reality where distinguishing between wants and needs isn't always clear. Since you've read this far, I deduce that you're here for the red pill. Welcome! Let's get critical with our purchasing decisions!

The first skill I'd teach you is detecting spending triggers. What are spending triggers? A spending trigger could be an *emotion* or a *circumstance* that causes you to want to spend money without prior planning. Can you think of any spending trigger associated with the iPhone? Okay, let me help out!

Social and peer pressure is one: Everyone seems to be using one. So, you join the iPhone gang. Ever heard someone say, "Android phones are just cheap plastic," or "iPhones are for rich people?" Well, that's fashion for you.

Then, there's the emotional part: The ecstasy of owning an iPhone like everyone else. Seriously, according to The New York Times, about 90% of teenagers choose iPhones over other brands (Mickle, 2023).

That's not all though. Sometimes, we spend impulsively when stressed—it's a kind of coping mechanism to escape stress.

Does this mean you shouldn't buy an iPhone or that item? Of course not, but it's crucial not to dig a debt hole for yourself. So, here's the million-dollar question: How does one make the right purchase? Before you click "add to cart," you might want to question your decision.

The iPhone 15 was just launched, and like everyone else, you're looking to get a hold of it. Wait a minute...

Do Some Research: Why are You Buying The Product?

Normally, you want to know what you're buying, right? The next course of action is to carefully research the new model. Go beyond the adverts. What's the most important feature of a mobile device to you? Gaming? Music? Social media? Videography?

I'm guessing it's more videography and photography. So, what do you do next? Well, check out the camera specs and compare them with your current device. Is the camera actually better? Is the processor better? Is the display better?

What are other people saying about the device? Who knows, it may have some drawbacks not mentioned in the adverts. Companies won't say, "Oh, this product has certain flaws," No! It's up to you to do your homework. If the tech language is too complex for you, you can check out tech folks who review mobile devices on YouTube or TikTok.

Consider the Value: Do I Really Need this?

Don't "add to cart" just yet! It feels cool adding another Nike shoe to your collection even though you already have one for every day of the week. The question to ask yourself here is, do you really need this? Why are you considering purchasing it in the first place? If you already have enough shoes, why buy another pair when that money could be saved for college?

What purpose does the purchase serve? Are you buying it because you need it or you simply fancy it? Don't forget, the companies know you might not need their products, but they make it seem like you do. Taking time to answer these questions will dim down the excitement clouding your judgment.

Did I Budget For this?

Spending outside your budget is your one-way ticket to debt-express. The rule is simple: If it wasn't in your budget, then don't buy it. Did you save and budget for a car worth $10,000 then saw an ad for another worth $12,000? At that moment, you may be thinking $2000 is just a little extra. But I tell you, that's a grave mistake! That's an unplanned $2000 spent!

What Will it Cost Me?

You've done your research? Still thinking of that product? Wait a minute. Have you considered the impact that purchase will have on your finances and lifestyle? You might miss the picture if you think of the immediate financial cost alone. For instance, some items may require constant maintenance, which means more expenses. The question is, will those expenses align with your goals? Will the expense affect your budget in the long run?

Always remember that advertisements are designed to compel you to buy products even if you don't need them. To sum it up, you should always think about the pros and the cons before making a purchase.

EMBRACING THE MONK LIFESTYLE

People often complain that living on a budget means depriving oneself of the things that bring happiness. But that's a misconception as old as T-Rex's fossil! Ever wondered if monks thought their lives unfulfilling? In reality, monks and people who adopt such lifestyles understand that living life on a budget is not about being cheap; it's about making mindful choices that align with their values.

The ability to spend consciously and make deliberate choices to ensure you're getting the best value of your money is called *frugal living*. You'd be surprised to find out that many wealthy people adopt such a lifestyle. For some, it was a key factor to achieving their wealth. One such individual is Ed Sheeran, the superstar singer-songwriter. Surprised? Don't be! Despite his immense success, Sheeran has stayed grounded when it comes to money, and there's plenty we can learn from his approach.

No Matter How Much You Make, Keep a Budget

Do you believe Sheeran actually gives himself an allowance? Despite being worth more than $200 million, he limited his spending to $1000 per month for a very long time. He once said, "I have enough to be comfortable, and the rest goes to help people." If you met him on the street, you wouldn't suspect him to be a celebrity—an international celebrity for that matter.

> *"Making more money won't resolve all your financial problems. Learning how to manage money correctly will. And frugality can help with it."*
>
> — *YIN MAYTHU*

Like Ed Sheeran, you need a budget. Think of it this way, assume your financial and life goals are a destination, and a budget is like a map to that destination. Being frugal isn't about starving yourself of life's pleasures; it's about prioritizing —spending on the essential things first.

A Lamborghini is Cool, But a Honda Will Get You There Too

If given the option, everyone would prefer a Lamborghini over a Honda. There's no shame in that. Who doesn't want to look cool? But here's the harsh truth: Reality doesn't always align with our desires. Many multi-millionaires and billionaires have paid a hefty price for their extravagant spending.

Take, for instance, Sam Bankman Fried, popularly known as SBF, who was once praised as the cryptocurrency messiah. Although the story on the news goes, "He was running a scam," it is actually a matter of inappropriate use of public funds, which was used to fund his extravagant lifestyle and that of his co-workers.

Overspending is like a rollercoaster ride without safety restraints. So, I implore you, before you set your heart on that Lamborghini, pause and ask yourself, "Can any other product serve the same purpose?" This thought process can apply to every purchase you make.

THE PHILOSOPHY OF SAVING

 "Everything is in excess except money; thereof, it should be well managed."

— *LAILAH GIFTY AKITA*

Saving transcends mere cutting costs on your daily expenses—it is a philosophy of living. The philosophy of saving is one adopted even by many creatures on the planet. For instance, the success of an ant colony is defined by how many supplies they've saved. This philosophy is one adopted by even the richest of folks.

Warren Buffet, the fourth richest man on earth, said, "Do not *save what is left after spending*, but *spend what is left after saving*." Take a moment to reflect on those words. How have you been saving? I can guess that you are used to saving what is left after spending. To save properly, you must view it as a way of life—a philosophy that weaves purpose, mindfulness, and participation into the fabric of your existence.

Like all endeavors of life, you need a savings plan. So, let's begin.

Creating a Savings Plan

Create a saving plan by following these steps:

1. What's your income total?

The first phase of your savings plan is having an overview of your finances. What's the total sum of money coming into your pockets? This includes your allowance, any income from part-time jobs, or even money you make from occasional side hustles. If you receive regular paychecks, look at your net income (Your net income is the amount you have after taxes).

2. Gather all of your expenses

Remember our budget sheet from Chapter 3? It's time to put it to use once more. Tracking down all your expenses can be hard, especially since unexpected expenses always arise. However, it helps to categorize your expenses. For instance, all costs relating to housing can be grouped into one. Other categories include food, transportation, medical expenses, entertainment, social activities, savings, and so on.

Don't forget to include all online transactions. This includes amounts used to purchase TikTok coins, gaming currencies, subscription fees, and so on.

3. Jot down your financial goals

Carefully go through your budget and expenses: Where have you been spending your money, and how much money is left for saving? Doing this tells you where you've been and where you're headed; it's like a map! Now, it's time for some introspection—ask yourself these key questions:

- Are your spending and saving choices aligned with your values?
- What's the debt situation? Do you have more debt than you're comfortable with?

Now, for the important part, what are you hoping to accomplish in life? Close your eyes and paint a mental picture of your ideal life. What goals do you hope to achieve in a few years to come, and in the distant future? Can your current financial habits and situation support those goals?

4. Plan to achieve your financial goals with budgeting

With your financial goals down on paper, it's time to figure out how to turn these dreams into reality. Take a closer look at your expenses, which you've already outlined in step two.

Considering the goals you've also laid out, ask yourself, are your finances on the right track? Are there areas where you can trim down expenses? Maybe you're spending too much on eating out or streaming services you hardly use. Identifying spending habits that will hinder your progress is crucial in this phase.

Are your expenses already as lean as they get? If so, will your current financial situation drive you to your goals? If not, then consider earning more money. You're left with four options:

- hunt for a better job
- gain a new skill that will lead to higher paying jobs
- find a part-time job to give your income a boost
- doesn't always work, but you could consider asking for a raise at work

There's also the option to rearrange how much of your money goes into different aspects of your life. That's where budgeting comes into play: There are many budgeting methods out there; for example, there's the 50/30/20 method.

50/30/20 Budgeting Method

This method divides your expenses into three categories: The first is needs, the second is wants, and the third is savings and debt repayment. With the 50/30/20 budgeting method, every category is allocated a certain percentage of your money; 50% for needs, 30% for wants, and 20% for savings and debt repayment. For instance, if your net income is $600, then $300 (50%) of that goes to needs, and so on. Get the point? Work this into your own plan. Will 50% cover your needs? If not, you can try the 60/20/20? Still not enough? Keep adjusting until it satisfies your goals.

5. Can you achieve your goals in your current situation?

Remember, the key is to save towards your goal. First, figure out how much you need to achieve it. For instance, if you aim to save $3,000 by putting away 20% ($120) of your $600 monthly income, you must calculate how long it will take. Simply divide $3,000 by $120, and you'll find it will take 25 months, which is equivalent to two years and one month. The essence of doing this is to know the timeframe needed to save, and this gives you room to plan ahead or adjust your strategy.

Remember, there isn't a one size fits all; some techniques may work better in certain situations and fail in others.

Overcoming Challenges of Saving

People struggle to save. That's a known fact. A startling survey carried out by *Bankrate.com* revealed something quite scary. From their survey, only 16% of the total population are able to save at least 15% of their income. Even worse, 21% of American adults don't have any savings (Dixon, 2019). Why? You see, the finance world is a complex one: It's akin to an adventure video game—the higher the stage, the more difficult it is. Each new stage comes with new challenges.

The question is, how do you overcome these challenges? Let's dive into the most prominent challenges and how you can overcome them.

The Relentless March of Inflation

Inflation refers to the rate at which your money loses value, and I must tell you; inflation is a relentless beast. It lurks just behind the bush waiting to pounce on you and everyone. Currently, the inflation rate of the U.S. sits at 3.7%. In other words, if a bag of chips cost $10 last year, you need an extra 37 cents to get it now. To put it simply, the cost of living is steadily increasing.

The only way to escape the clutches of inflation is to earn more money, not save more. There's a twist though; jobs aren't paying more nowadays. So, what option do you have left? That'd be investment. Saving may provide a safe haven for your money, but it can't keep up the pace with inflation. It'll eventually become worthless as time goes by. That's why investing is the viable option.

Not Knowing How to Save

The lack of financial literacy still persists. Many people haven't picked up this book like you did and don't know how to create a budget, save money, or invest. So, people live their lives without a plan, without a vision of the future. Plus, the lack of knowledge about finances breeds fear—fear of making bad financial decisions. But that's not you! You've learned how to make your budget and construct saving techniques to safely get you to the destination where your dreams come true!

Debt and Unexpected Expenses

You can't save if there are debts to pay. Already, the average American is knee-deep in debt: $986 billion in credit card debts, $11.92 trillion in mortgage debts, $1.55 trillion in vehicle loans debt, and $1.60 trillion in student loan debt (Fay, 2022).

How does one fight debt on one front, inflation on another, and savings on yet another? The key lies in budgeting and the application of debt repayment and saving plans. Like Sun Tzu said, "The general who wins the battle makes many calculations in his temple before the battle is fought. The general who loses makes but few calculations beforehand" (AZ Quotes, n.d.). Adulthood is a battlefield, and you cannot win without a plan.

ACTIVITY 3

We've talked a lot about creating a savings plan. Now, let's show you what a saving plan looks like and how you can create your very own!

Use this template to create your own saving plan. Cheers!

Savings goals		
Emergency funds		
Vacation		
New laptop		

Income tracker			
Income		Amount ($)	
Job income			
Side gig			
Saving Plan			
Categories	Monthly Budget ($)	Actual Expenses ($)	Amount saved ($)
Housing			
Utilities			
Groceries			
Transportation			
Dining out			
Entertainment			
Miscellaneous			

END OF SESSION

In Chapter 5, we've delved into the fascinating world of smart shopping and financial savvy. We've uncovered the tricks our minds play on us while shopping and how advertisements can lead us astray. But here's the game-changer: With awareness, we can resist these temptations.

Research becomes our ally, allowing us to see beyond flashy ads and grasp a product's real worth. Asking key questions, like whether we truly need an item or if it aligns with our goals, helps us beat impulsive spending.

Budgeting becomes our trusty guide, offering various methods to suit our financial needs. Saving becomes a lifestyle, not one of deprivation, but of mindful choices in line with our values.

Now, it's time to act. Analyze your spending triggers, craft a budget, and create a savings plan. Challenges will arise, from inflation to emotional triggers and unexpected expenses. But with strategies like "paying yourself first," technology-assisted banking, and celebrating milestones, you'll stay on track.

Now that you've mastered the art of shopping smart and saving, let's prepare for unexpected twists. The next chapter will guide you through emergencies and insurance.

6

EMERGENCIES AND INSURANCE REFERENCES

L ife is full of surprises, and not all of them come with a gift. There were times it seemed every step I took back-fired—the car broke down, sudden illness struck, or expenses loomed. I can't even fathom what I would have done without the support from my family.

Think about what you would do if your car broke down tomorrow, or if you suddenly needed urgent medical care. Are you financially prepared for unexpected emergencies? So far, we've discussed creating a plan A; now, it's time to emphasize the importance of having plan B.

PREPARING FOR THE UNEXPECTED

In an earlier section, we talked about setting some money aside for savings. But savings usually have a purpose: Helping you achieve your life goals. However, emergency funds have a different purpose. They act as your last resort. Just like in *Avengers: Endgame*, when all hope seemed lost, the Avengers turned to a last-ditch plan, which eventually saved the day. Similarly, in the world of finance, having a plan B is equally essential.

An emergency fund is your safety net, your "get back on my feet" fund, for when life's crises come crashing through the front door. As a general rule of thumb, your emergency fund should cover your expenses for at least six months. Another general guideline with keeping emergency funds is to save them in an account that you can access only in case of an emergency. So, stashing your emergency fund in a fixed deposit account wouldn't be the wisest choice, would it?

BUILDING YOUR OWN EMERGENCY FUND ACCOUNT

Remember, an emergency fund isn't just a financial cushion; it's your shield against life's unexpected challenges. And trust me; they are bound to come. Creating your own financial safety net begins with building your own emergency fund account. Let's jump right in and discover how to build and grow this crucial resource.

Step 1: Open a savings account that you can easily access in times of emergency. It's crucial that this is a separate account dedicated to just savings. Can you guess why? Will it help you resist the temptation of spending on non-emergencies? Right!

Step 2: Set up an automatic transfer from your main account to your emergency fund account. Not sure how much to save yet? Remember we talked about budgeting methods? Use the knowledge to figure out what percentage of your income to save. Automating your savings ensures you set money aside without having to even think about it. It's another great way to avoid spending excessively.

Step 3: You want to make sure you're steadily saving or increasing the amount you save until you've saved enough funds for at least six month's expenses.

Step 4: Explore income-boosting opportunities. What's the situation with your budget? Can you save? Will you be able to save consistently? Increasing your income can significantly boost your saving potential. But remember to increase the percentage you save as your income increases.

You'd feel much safer knowing you've got an emergency backup plan. In addition, you get a significant confidence booster when your finances are sorted out.

INSURANCE 101

Insurance sort of works like an emergency fund, only you're not saving in your bank account. Financial institutions came up with a brilliant idea. They thought, "Hey, people are finding it difficult to save for emergencies. Why don't we create a special account and help them manage their money?"

The purpose of having an emergency fund is to protect yourself from uncertainties, right? With insurance, not only are your funds managed by a financial institution, but the purpose of the insurance fund is specific. For instance, insurance could be a backup plan against health risks, property damage, and even loss of life.

IMPORTANT INSURANCE TERMINOLOGY

These are insurance terms I think you should know:

- **Premium:** The premium is a monthly or annual fee, like a subscription fee, you pay to your insurance company. Paying this fee regularly keeps your insurance coverage active.
- **Deductible**: Sometimes, the insurance company doesn't cover all the costs and requires that you make an initial payment to fix your problem before your insurance benefits kick in.
- **Policy**: An insurance policy is the contract or agreement between you and the insurance company. It explains what your insurance covers, what it doesn't cover, and the cost of the premium. It tells you what you get out of your insurance package!
- **Coverage**: Coverage is the term that tells you what your insurance protects you against. For instance, if you have car insurance, it means the coverage includes only damages to your car, when your car gets stolen, or other problems you encounter on the road.

- **Beneficiary**: A beneficiary is the person or group you choose to receive the money from your insurance if something happens to you. They benefit from your insurance policy, hence the name "beneficiary."
- **Claim:** A claim is a formal request you make to your insurance company when you need their help.

Types of Insurance

There are a ton of insurance plans out there offered by different insurance companies, but the most common include:

1. **Life Insurance**: Life insurance is an agreement you make with a company to help your loved ones if you pass away. A lump sum of money is paid to your named beneficiaries when you pass away.
2. **Health Insurance**: With health insurance, the insurance companies help with your medical bills when you get sick or injured. It can include a doctor visit, hospital stays, and medications.
3. **Auto Insurance**: While health insurance protects your health, auto insurance protects your car. Like other insurances, you pay the premium, and if your car gets damaged in an accident or stolen, the insurance company steps in to help you pay for your repairs or get a replacement. Cool, right?
4. **Long-term Disability Coverage**: This type of insurance covers you in a situation where you get severely injured or an injury keeps you from working

for long periods. This is helpful since, during the disability period, you'd be unable to work. If you're working a dangerous job, this will be a good fit. The insurance company supports you while you recover.

LIFE IS UNPREDICTABLE: SO IS DEATH

Every day, we hear about unusual deaths. Sometimes, it's the result of a chaotic highway accident, and other times, it's as unexpected as having a stroke while sleeping. But what I heard on the news some time ago made me realize that death was more unpredictable than I thought.

Death rarely crosses our minds, especially when we're feeling healthy and unstoppable. We all believe we have a long, bright future ahead, and that's a wonderful outlook to have.

However, the victim of this story had the same outlook on life. It didn't take much for everything to change dramatically.

According to the news, a Brazilian man passed away in his sleep. Now, that doesn't sound too strange, right? But here's where it gets truly bizarre. While he was peacefully sleeping next to his wife, a massive cow crashed through the roof (Quinn, 2013). Yes, you read that right, a 6000-pound cow!

Initially, he seemed fine, and after removing the cow, they went back to sleep. However, things took a sudden turn. The impact from the cow had damaged the man's internal organs, though he didn't realize this at the time. By the next day, he was tragically pronounced dead. It's a heartbreaking tale to tell.

Upon reading this news, I couldn't help but wonder why he didn't go to the hospital. It dawned on me; he probably thought of the cost. Since he wasn't experiencing any immediate pain, he likely saw no need to incur unnecessary expenses. But consider this: What kind of insurance could have helped when the victim's house was damaged? Suppose he hadn't died immediately; what insurance might have been beneficial if he had sought medical attention? And after his passing, what insurance could have eased the burden on his family?

Life doesn't always go as planned, but with the right insurance, you can be prepared for whatever life throws your way.

WHEN INSURANCE SAVES THE DAY!

While some people try to live without giving a second thought to the uncertainties of life, others are well-prepared for it. Just the other day, I was doom scrolling through my LinkedIn (Yes, we old folks do that too!) feed one day when I stumbled upon a story posted by a commercial insurance strategist. It was about an optometrist who owned a building in a pricy neighborhood. Someday, there was an accident, a drunk driver lost control of their car and rammed into the building of the optometrist, causing a lot of damage (Lipson, 2020).

All this happened with the optometrist away from home and attending to his job. I guess he must have been in shock when he got the news. Usually, the person at fault fixes the damages they caused, but the drunk driver didn't have the money.

Although the drunk driver had insurance, it wasn't enough to cover the cost. So, who came to the rescue? The insurance company of the optometrist!

CHOOSING THE RIGHT INSURANCE

The same way you investigate the specs of a gadget you intend to buy, is the same way you have to investigate insurance coverage and policies. It can get tricky choosing the right insurance package because it's all about warding off risk—risk that you're uncertain of. Before you choose any insurance, consider these factors:

What's Really At Risk?

Let's assume you work out every day, eat your carbs, go for regular checkups, and do everything to stay healthy. Health insurance may not be of much importance to you.

Suppose you also drive a few miles to work each day, and your car is an important asset. In such situations, which insurance do you think you'd be more likely to have?

What insurance you need depends on your lifestyle and priorities; you may lean toward one type of insurance over the other.

Cost

Money is always an important aspect of anything, right? As you now know, insurance isn't free. It comes with a premium. So, the question to ask yourself is, "Can I afford this?" The cost of the insurance also depends on the type you want.

You might not be able to afford all insurances together, so figuring out where you're most at risk is key. The good news is, there are lots of insurance companies out there that offer a range of affordable packages. So, you've got your homework cut out for you.

Coverage limits

Suppose you have a video game console worth $500, and it gets stolen. If your insurance only covers up to $300 for electronics, you'd be out of luck for the rest. Hence, knowing how many damages the insurance covers is a must: Ensure the coverage matches the value of your stuff.

Deductibles

Typically, higher deductibles mean you pay lower premiums, but it means your emergency funds must be stacked up. Do you see where your emergency fund and insurance work together? You can also opt for higher or lower deductibles and pay less out-of-pocket costs.

Customer Service

How an insurance company treats its clients is important. You can read through the reviews left by other clients about the services of the insurance company. You'd be speaking to someone on the other end of the phone when you're making a claim on your insurance benefits. It'll be much better if the voice on the other end was warm and welcoming.

Extra Perks

The insurance industry is a competitive one. Some insurance companies offer extra perks to attract more clients; for example, you could get roadside assistance as a bonus on your auto insurance.

Don't forget that insurance is a means of getting yourself ready for unexpected downfalls. It's like your bodyguard, so be sure to ask a lot of questions, and choose the insurance that makes you feel safe and secure.

ACTIVITY 4: BUILDING AN EMERGENCY PLAN

1. Determining the right insurance coverage for your unique circumstances is a critical step in financial preparedness. To help you assess your insurance needs effectively, consider the following questions:

- Do you have dependents?
- What assets do you want to protect?
- Are you the primary income earner in your household?
- Do you have substantial savings and investments?
- What is your health status?
- Do you have valuable personal items or collections?
- Are you planning for retirement?
- Have you evaluated your existing policies recently?

2. Not all banks or financial institutions are created equal; that's why it's crucial that you choose one that aligns with your goals. Follow these prompts to help you research and compare different type of banks:

- What is most important to you in a savings account? Consider the interest rates, fees, accessibility, and other features.
- Start your research. The internet is your playground.
- Plot a table or comparison chart that compares essential factors like interest rates, fees, and so on of the banks you're considering.
- Read reviews and ask for recommendations from experts, friends, or family.
- Take a tour of your bank. Explore the bank branches or their website to gather detailed information.
- Finally, make your choice.

END OF SESSION

In this chapter, we've explored the critical aspects of emergency funds and insurance to prepare you for life's unexpected twists. Life is unpredictable, and unforeseen challenges can arise at any time. These surprises may include car breakdowns, sudden illnesses, or unexpected expenses.

Unlike regular savings for life goals, emergency funds serve a different purpose—they act as a plan B. Emergency funds should also not be kept in your main account. They're strictly for periods of crisis. Experts recommend having an emergency fund that covers at least six months of living expenses.

Although emergency funds are great, they can be difficult to come up with or manage. Financial institutions solve this problem by coming up with insurance. Insurance adds another layer of protection against life's uncertainties. It involves paying premiums to a company in exchange for financial coverage in specific events. It's crucial that you understand what you're signing up for and ensure you're getting the best deal.

Now that you've mastered the art of preparing for the unexpected, let's explore ways you can generate wealth.

7

PLANNING FOR COLLEGE AND BEYOND

Did you know that individuals with a college degree tend to earn at least $1 million more than people without one (Elkins, 2019)? It's a remarkable statistic that underscores the immense value of higher education. However, there's a rising trend where young folks like yourself are beginning to doubt the value of education. But in a world where young minds like yours are increasingly questioning the worth of a college degree, it's time to dig deeper.

Just a few years ago, I met a college graduate named Sarah. She had invested time, money, and energy into her education, only to find herself uncertain about her career prospects. Sarah's story isn't unique. Many students are troubled by the fear that they've wasted their resources on a college degree. This brings up the question: Is there a disconnect between education and employment?

That's not the case either because the unemployment rate for 2023 sits at 3.50% at the time of writing (Ycharts, n.d.). But 39% of college dropouts say it wasn't worth spending on (Gilbert, 2020). This raises yet another question; why are Gen Z losing faith in education? According to The Quo Education Pulse Survey, the root often lies in students not receiving proper guidance when choosing their career path (Burt, 2022).

That's precisely why we're here today. In this chapter, we will explore strategies for selecting the right college, discovering ways to secure funding for your education, crafting a plan, and choosing the career path that best aligns with your dreams and financial goals in life. College is the gateway into a complex world, and with the right guidance, you'll be well-prepared to thrive.

THE COLLEGE FINANCIAL JOURNEY BEGINS

Remember those childhood dreams we all had? Becoming doctors, actors, singers, or dancers? We make this decision when we know little to nothing about the dynamics of the economy.

Over the course of my career, I've witnessed young lads who realize that their dreams weren't anchored in the reality of the world around us. It's like gearing up for a marathon only to realize you're on the wrong track.

"This career path isn't fulfilling."

"I spent so much money in college only to earn peanuts!"

"Just found out the course I'm taking at my college wasn't even accredited."

"It's tough finding a job because my school has a bad reputation."

These are just some of the hundreds of complaints I've heard over the years.

The number one advice I give those I mentor in the finance industry is *"always choose a career path that isn't just about following your heart; it's about aligning your dreams with the heart-beat of the economy."* The same goes for college. There are many factors to consider before filling up that application form. Let's take a look at them:

Aligning Your Passion with College Offerings

There are colleges that offer a wide range of courses, and then there are those that are highly specialized. Take, for example, you have a passion for marine biology, and your dream is to dive into the depths of the Pacific Ocean. A specialized college like the Scripps Institution of Oceanography in California or the Rosenstiel School of Marine and Atmospheric Science at the University of Miami will be better options. The thing is specialized colleges often excel in providing tailored resources and state-of-the-art facilities that are precisely suited for your field of interest. This ensures that your educational investment yields the best possible returns.

Cost and Fees

Next on your list should be none other than the tuition fees and cost of schooling. Remember, we discussed the cost of education and the rising student loan debts? Now, if you spend $40,000 or have its equivalent in debt, you'd expect to have a job that can give you a return on your investment in a year or two, right? Ask yourself these questions:

- Are the fees too high, and if so, do they offer scholarships?
- Is the cost of their education worth its market value?
- If school A offers the same program at a cheaper cost compared to school B, then school A would be the better option, right?"

Compare the prices of different schools; you can do this the old way by plotting a table and inputting the tuition fees of each school.

College Campus Environment

Although colleges are centered around academic activities, the campus and environment have a significant impact on your experiences as a student and your quality of life. Look beyond just the classrooms and halls. Is the campus vibrant and inviting? Are there recreational facilities and places to relax and socialize? What kind of drinks can you have? What are the rules and regulations? Will you feel welcome in the campus community? Are there dress ethics? And so on.

Another aspect to consider is access to resources. Are there research facilities, libraries, or other specialized resources necessary for reaching your academic goals? Don't forget the cultural and social activities of the surrounding environment.

You want to avoid attending a college where you feel alienated or always the 'odd' one out. That's why you ought to consider the diversity of the campus and its surrounding community.

Then, this is very important, choosing a college in a city with a thriving job market increases your access to internships and job-preparation opportunities. This can provide a very rich network, which is essential in today's job market.

Reputation and Rankings

During conversations with your friends, or online friends, you'd certainly come across various rankings and hear about the reputation of some institutions. We humans always ascribe a hierarchy to a lot of our economic activities, and college isn't left out. Typically, the higher the ranking the more prestigious the college. Some organizations, like the Times Higher Education, QS World University Rankings, use various criteria to assess colleges. For instance, a college may be ranked based on research output, student-to-faculty ratio, and so on. Visit *www.timeshighereducation.com/world-university-*

rankings/2023/world-ranking to check the ranking of your college of interest.

There are some helpful tools you could also use to help assess your career prospects and choice of college. For example, there's College Board's BigFuture that allows you to target colleges based on your career goals and interests. CollegeData is another free tool that helps you compile a list of colleges offering scholarships. Other tools include College Scorecard, Niche, CollegeXpress, and so on.

HAVEN'T SAVED ENOUGH COLLEGE FUNDS?

Let's face it; the cost of education can hit you like a ton of bricks. We're talking about shelling out $32,637 each year just to obtain an undergraduate degree from a public university (Hanson, 2023). And that's the average! It's enough to send any teen's wallet into hiding.

We can clearly see the effect of the increasing cost of education; I mean, student debts amount to trillions of dollars! When savings don't cut it, students reach out to the government for a bail out—student loans. We haven't begun to talk about living expenses.

As things are, the only option is to turn to financial aid. College financial aid offers you the opportunity to enroll for higher education and mostly cover your educational expenses.

The first financial aid bus stop most students run to is private or federal student loans. But let's not forget, loans don't come cheap. For most people, it's a burden they'd bear for a very long time. I really don't want that to be you! So, fear not; there are other alternatives that you can use to fund your education and not have to worry about paying back. Yep! I'm talking about:

- Scholarships: A scholarship is a financial award given to a student to help them pay for their education. They are usually merit-based, meaning you need to show excellence in your academics, athletics, or other criteria.
- Grants: Unlike scholarships, grants are awarded based on the student's financial need. They also could be provided by government or private institutions. Additionally, they are not as common as scholarships and often provide lesser amounts than scholarships.
- Work-study Programs: These are programs that allow students to work part-time jobs, often on campus, so they can earn money to pay for their educational expenses. With this program, you can earn at least $7.25 per hour.

These are basically "free money." However, because they are limited, you need to prove that you deserve them.

One reason many fear to apply is because of the rigorous application and screening involved. However, nothing good comes free, right?

I recall reading about a teen like yourself who got accepted into, drum roll, over 50 colleges and universities. You heard that right! She earned over $1.3 million in scholarship money! How did she do it? Good question. The straightforward answer is, it takes practice and lots of wits!

WHERE TO FIND GOOD FINANCIAL AIDS

If you're looking for federal grants, work-study, or loans, visit the U.S Department of Education's website where all the good stuff is hidden in plain sight. Proceed to fill the Free Application for Federal Student Aid (FAFSA). The FAFSA form consists of about 100 questions that you need to answer. The application process begins every 1st of October; you don't want to miss the deadline too. The website address is *https://studentaid.gov/h/apply-for-aid/fafsa.*

For non-federal financial aid, you'd need to do a little more digging! People often miss these opportunities because they are more difficult to find. The first place to enquire from is the college itself. Visit their website and search for scholarship openings. Like I said earlier; these are based on merit, so be prepared to write essays or tests, attend interrogative interviews, and perform other tasks.

The U.S Department of Labor also provides you with a free scholarship search tool. You can access that from *www.careeron estop.org/toolkit/training/find-scholarships.aspx.* You can filter the search, which contains a database of more than 8000 awards.

You have a very powerful tool at your disposal, I mean the internet. You could also try Google searching for grants and scholarship offers online. Good luck!

ENERGIZE YOUR MOTIVATION WITH COLLEGE VISITS!

Occasionally paying a visit to your campus of choice will give you a morale boost to chase your educational goals. It isn't just about checking out the physical location: It's an opportunity to ignite and sustain your motivation toward achieving your educational goals.

First thing first, how do you get in? Your high school may have cool programs that arrange group trips to colleges. Visiting your buddies gives the event an extra thrill! You could also plan a road trip with friends and family and visit campuses by car or public transportation. Find out from the college if they have any terms you must meet before paying a visit.

Campuses are usually large hectares of land. Arm yourself with some equipment—a compass, telescope, thick boots, just to name a couple necessities... Just kidding! It's not a safari tour. Before you set off, contact the school's admission office and inform them of your visit. Schedule a guided tour; they know all the shortcuts and fun spots. If they allow you to walk around yourself, that'll be awesome too.

On your trip, ensure to check out the cool places; explore the library, check out the sports center, peep into a few classrooms if possible. Don't forget to jot down any experience. What were the exciting places? Do you fancy any buildings? Take notes of it all!

ACTIVITY 7: A COLLEGE PLANNING WORKSHEET

Research each factor that will help you determine which college to apply for. Take your time to research what each college has to offer and fill your findings in the description column. Repeat this for the top three colleges you wish to attend.

Priority	Factor	Description
1	Research opportunities	
2	Academic programs	
3	Cost and fees	
4	Location	
5	Social environment	
6	Reputation	

With college planning under control, let's look forward to the other major life transitions and the financial strategies that will support your journey.

8

LIFE TRANSITIONS AND FUTURE PLANNING

As you toss your graduation cap into the air, you're not just closing the college chapter; you're stepping onto the grand stage of adulthood. While you've gained freedom, it comes bundled with responsibility, and the decisions you've sown earlier will start to bear fruit.

I vividly recall my own college graduation, as if it were yesterday. The thrill, the bittersweet mix of farewells and new beginnings, the knowledge that I'd conquered one phase but might not see some of my friends again. I remember it all. School was officially behind me, and the real world kept knocking on my door.

Adulthood made college look like child's play. I soon discovered that there were many lessons I had yet to learn and that only the real world could teach me.

I wasn't truly independent though; the job search had only begun!

GETTING A HANG OF THE JOB HUNT AND SALARY GAME

"Investing in yourself is the best investment you will ever make."

— *ROBIN SHARMA*

Having a college or high school certificate is essential for landing a job, but guess what? It's not the only thing you'll need to succeed. These 'extras' are the skills they don't teach you in school, and I bet you didn't learn them in your class-room either! The first skill you need to add to your arsenal is mastering the art of applying for jobs. You'll find these tips helpful when searching for a job. They include:

- **Know yourself**: Before you begin sending in those resumes, take some time to reflect on your path ahead. Do you have the skills required? What are your strengths and weaknesses? What is your personality type? These are essential questions you must ask because they help narrow down your career choices and increase your chances of nailing a job. To know the skills you need to successfully land your dream job, study the job requirements and descriptions.

- **Make a company list**: Gather a list of all the companies that you're capable of working with. Doing background research is easy these days, thanks to the internet. Find out what the company culture is, salary range, job description, and so on. Carefully study the job description because it'll be your lifeline when writing your resume and cover letter. You could put this in a spreadsheet; keeping track of jobs.

- **Write one resume per job**: Suppose you're applying to different industries, including to a posting for a social media content creator at Netflix or a social media content creator at Apple. You shouldn't use the same resume. Why? Because each job most likely used different keywords in their job description.

- **Build your brand**: Today's job searches and recruitment events take place online. You want to have an online profile (preferably LinkedIn) that showcases your passion and skills. Optimize your online profile for industries you're interested in, and this will make you more visible to potential employers. Now, do the same for multiple job search sites; *Monster.com, Job.com, Glassdoor.com, and Indeed.com.*

- **Begin to apply**: Now, it's time to put your efforts to the test: Send emails, containing your resume and cover letter, to the companies you've selected. If you don't get responses, repeat the steps and try again. While you wait, begin to prepare for the interview. Good luck!

Negotiating Your Salary

Job postings usually come with a specified salary or salary range. However, sometimes, the table is open for negotiation. Bear in mind that the goal of the human resource manager is to employ someone who'd benefit the company.

Working culture is improving, but not all companies are improving at the same speed. Most companies always put themselves first; hence, you need to put up a little 'fight' if you want to get paid what you deserve.

However, you also don't want to appear as if money is what's most important to you. Unless you don't want to hear from them again, you need to come up with a negotiation strategy.

- The internet is your friend, once again. Figure out the average industry salary for your job to set your salary expectations. Remember, the salary you settle with should align with your financial goals.

- During the interview, the company has the upper hand, so be careful not to use the wrong choice of words. "I'm not comfortable with the salary specifications," can cut your interview short. Typically, the interviewer asks for your salary expectations. Then, you can tell them why you think you deserve to earn such an amount. Your skillset, experience, living expenses, and industry standards, are some valid arguments that you can point out to the interviewer.
- If you're not asked for salary expectations, wait for when the employer asks questions like, "Do you have any questions for me?" "Are you satisfied with the working conditions?" and so on.

When you're just starting out, negotiating your salary might be tough because the job market is saturated with skillful individuals who'd take your spot even at a lower pay. Should you accept the job if the employer disagrees on your salary expectations? Before you say, "No," ask about other benefits offered by the company; for example, paid leave, time frame for salary reassessment, and so on.

If it's still not working out, taking the job depends entirely on your financial needs and goals. If the proposed budget is enough to cover your expenses, how about you give it a try to gain experience? This will give you better leverage in your next interview.

BUILDING A FOUNDATION FOR ADULT FINANCIAL RESPONSIBILITIES

Did you know that as many as 64% of Americans live paycheck to paycheck (Dickler, 2022)? Surprisingly, this includes individuals with jobs who earn substantial incomes, often referred to by economists as the "wealthy hand-to-mouth." Shocking, right? This raises a critical question: Are these individuals spending all their money?

While you may celebrate your first job and the steady income it provides, it doesn't all signify the end of your financial concerns. It's only a bus stop on a very long journey with many more bus stops ahead.

Soon, you'll come to find that more financial responsibilities creep in: These include finding your own place, providing for your family, homeownership, insurance, and much more. Each of these responsibilities plays a critical role in shaping your financial well-being.

So, how do you tackle the new line up of financial burden? Well, you'd have to go back to stage one: Making a list of your financial goals. Remember that your financial strategy will be designed based on your goals.

SAVING FOR RENT

Rent is a major milestone achievement and may eat up a large chunk of your income pie. According to Statista, the average cost of a one-bedroom apartment is $1,152 per-month. If you're renting in a state like California, you need to earn at least $37 per-hour to afford a single bedroom (Statistica, 2020).

That's a lot compared to a state like Arkansas, where the average rent is three times less. Even then, we can only assume the prices are going to increase as inflation kicks in.

Gosh! How can you afford to cough out that much when you're only just starting out? Consider the following tips:

- Instead of opting for a one-bedroom apartment, consider a two-bedroom apartment with a roommate. This way, you can split the rent, making it more affordable.
- Discuss the possibility of an extended lease with your landlord. They may be willing to lower the rent if they see you commit for an extended period.
- Choose apartments with amenities that align with your lifestyle. For instance, if you don't own a car yet, there's no need to pay for an apartment with a parking space.

- Winter periods are bad business seasons for landlords, so they slash down prices to attract tenants. When life gives you lemons, make lemonade.
- Unlike corporate rentals, private landlords have more control over their prices and are more flexible with it. Hope you have a solid negotiation game!
- If it doesn't significantly increase transportation costs or compromise your safety, seek out neighborhoods with lower demand.

Don't forget the ultimate strategy to raise money for rent: Creating a super budget that keeps your expenses in check, allowing you to allocate more money toward your rent.

FINANCIAL PLANNING FOR PARTNERSHIPS, WEDDINGS, AND CHILDREN

Here's another interesting fact you should know: Over 50% of Americans hide their finances from their spouse (Rogers, 2022). Surprisingly, people have yet to understand the need for financial transparency in their relationships.

Why should you discuss finances with your spouse? The truth is, your relationship status greatly affects your finances, and not being open about finances can cause conflicts in the home.

A survey by Finmasters asked 1150 people about their relationships and finances; they found that about 35% of the participants said discussing their finances with their spouse increased intimacy and satisfaction in the relationship (Rogers, 2022).

On the flip side of the coin, over 90% reported that finances caused conflicts in their homes. The consequence of such behavior? Can you guess what one of the leading causes of divorce is? It's money issues (Institute for Divorce and Financial Analysts, n.d.). But before you say, "Yes, I do," there's a lot of preparation to do.

And the Two Become One...

 "A goal without a plan is just a wish."

— *ANTOINE DE SAINT-EXUPÉRY*

Marriage is another bus stop where you need to set new plans and make different goals. But this time around, "My goals are..." becomes "Our goals are..." Your new map should have room for your spouse and children. You must realize that when you get married, your financial decisions affect the whole family.

This means that you cannot make these decisions on your own. With your spouse in the picture, you'll need to consider:

- Wedding Expenses: This is determined by the type of wedding ceremony. Do you need to skip the wedding reception? How many guests should attend?
- Number of children to have: It's crucial to bear in mind that the cost of raising kids has increased and might not be the same as that of your parents. According to one study, the annual

average cost of raising children is about $13,000 (Felton, 2023).

- Do you have enough knowledge to strategize? Consider a financial expert's advice. Most parents don't consider the advice of a financial expert before choosing to get married and raise children. A financial expert will have better insights and will be able to advise you on how to better manage your household.
- Your lifestyle difference: Living expenses are determined by lifestyle. One person may have a more frugal outlook on life, while the other may have a more luxurious outlook. In such a situation, both parties need to compromise or meet in between. Recall, it's one of the reasons couples fight over finances.
- Responsibilities: Relationships come with responsibilities. So, you need to figure out who does what. How do you split the bill? What about during pregnancy? Can the income of one person sustain the family?

Not all dates should be about regular talk. Plan some dates to strictly discuss both your finances. Both of you need to sit together to plan out the following details: The checklist below highlights crucial information you should discuss.

Financial information	Expense	Legal Considerations
• Tax filings and refunds • Credit card scores • Student loan debt • Income • Credit card debts • Bill-paying • Keeping up with accounts, • Statement filing • Investment goals • Retirement goals • Assets • Liabilities • Investment strategies • Financial goals • Career goals • Savings goals • Emergency fund • Net worth	• Household • Transportation • Clothing • Feeding • Childcare • Laundry and cleaning • Entertainment • Education • Children's education • Financial support to parents • Mortgage payments	• Divorce settlements • Cohabitation agreements • Records of inheritance • Marriage certificate

A LIFETIME OF FINANCIAL SUCCESS

Practicing good financial habits should be as regular as leveling up your favorite video game character. Just like in your games, your financial plan needs constant upgrades. We call this "financial reassessment," and it's crucial, especially when you enter a new phase of life.

It could be that you're:

- getting married
- having children
- Experiencing a change in lifestyle and goals

All these are factors that should trigger you to check your financial game plan.

So, to perform financial reassessment, follow the step-by-step guide below.

The first step involves checking the following components of your finances. Your:

- income
- expenses
- liabilities and debts
- assets

This will provide you with a clear understanding of your financial situation and guide you to make the right decisions.

Next, reflect on past choices. For example, did you dip into your emergency fund for a vacation? What were the consequences? This helps you learn from your financial history.

Ask yourself whether there are goals that no longer make sense to you. Maybe you had a goal to save for a new car, but now, you're prioritizing paying off student loans. Adjust your goals to match your current circumstances.

Thirdly, consider the risks of your new financial strategy or goals. Will taking on a new goal affect your emergency fund? What if your income fluctuates unexpectedly? Always have a backup plan and avoid risking more than you can afford to lose.

Now, it's time for you to update your finance plan. Modify your spreadsheets or journal to reflect the changes you've made and give it a final review.

If you are having a hard time figuring all this out yourself, especially when you're uncertain of economic circumstances, don't hesitate to seek the guidance of a financial advisor.

MAKING THE RETIREMENT DREAMS COME TRUE

During one of my doom scrolling sessions on social media, I stumbled upon a rather intriguing headline from Reddit: "I Feel So Lost And Stupid for Having No Retirement Savings." Initially, I assumed it was the cry of someone in their golden years realizing financial mistakes, but I couldn't have been more wrong.

It turned out to be a 30-year-old individual earning 30k annually, and they already had an impressive $90,000 saved up. Imagine being 30, earning $30,000 a year, and having saved an impressive $90,000.

You might think that's a lot, but there's a catch. Inflation is around the corner. When you consider inflation, you'll understand why the individual was very worried; their future wasn't secured.

That's why you need a retirement account. But, first, you must paint a picture of how you want life to be for you after retirement. Then, year after year, you manifest those dreams.

How much do you think this dream will cost? How do you know what to save per year? There are several factors to consider; these include the age you plan to retire, how much you're able to save before you retire, and your current age (very important).

RETIREMENT ACCOUNTS

Pay attention to this section because a retirement account is going to be your best friend. The retirement account is a special savings account designed for long-term goals, specifically retirement. There are many retirement accounts out there: For example, there's the 401(k), Traditional Individual Retirement Account, and Roth Individual Retirement Account, among many others.

Each investment account comes with special features; the major differences are the maximum amount you're allowed to contribute per year, the tax benefits, and so on. Using the Roth account, I'll show you how investment accounts work. Grab a doughnut and some soda; it's about to get technical.

Roth Individual Retirement Account

A Roth Individual Retirement Account (Roth IRA) is an investment account that allows you to save for your future while allowing you to enjoy certain tax advantages. Here's a breakdown of its important components, features, benefits, and potential drawbacks.

Components

The Roth IRA, as well as other retirement investments accounts, have the following components (Vanguard, n.d.):

1. Contributions: This is the amount of money you can pay into your investment account. For the Roth IRA, the maximum amount of money you can contribute yearly, if you're less than 50 years of age, is $6,500.
2. Investment vehicles: The Roth IRAs allow a wide range of investment vehicles like stocks, bonds, mutual funds, and more.
3. Withdrawals: This is the amount of money you can withdraw from the investment account every year.
4. Minimum distributions: This is the amount you must withdraw from your investment account once you reach a certain age. It varies from one investment account to another.

Benefits of the Roth IRA

The unique features that separate the Roth IRA from others include:

1. Tax-free withdrawals: The Roth IRA offers a tax advantage since no taxes are applied on the money saved in the account. Other IRAs don't offer this advantage; instead, each withdrawal you make is taxed.

2. Flexibility: Unlike some other retirement accounts, Roth IRAs don't have required minimum distributions (RMDs) during the account holder's lifetime. This means you can let your money grow tax-free for as long as you like.

3. Long-term savings: Because you can withdraw whenever you like, you can allow your savings to grow, which results in an even larger retirement nest egg.

The Cons of the Roth IRA

All retirement accounts come with setbacks. You have to do your homework to see how the setbacks affect your goals. The disadvantages of the Roth IRA include:

1. Income Limits: You can't ride the roller coaster if you aren't tall enough, right? With the Roth IRA, if your income is beyond a certain limit, you may not be eligible to enjoy its benefits.

2. No immediate Tax Deductions: Which do you think will grow more trees: A pot with five seeds or a pot with one seed? The one with five seeds. The Roth IRA advantage is also a double edge sword. Since your savings were taxed before you made a deposit, it means you have less money to contribute to the account. Hence, less money trees grow.

3. Penalties for Early Withdrawal of Earnings: Because retirement accounts are meant for retirement (Duhh!),

withdrawing money before you retire (or reach age 59.5) attracts a fine.

Which investment account to use depends totally on your retirement goals. You have to compare the benefits against the cons and choose those which will drive you to your goals faster, and most importantly, with little risk.

SAVING FOR HOMEOWNERSHIP AND FUTURE BIG-TICKET EXPENSES

We are not leaving any stone unturned. Now, let's lay out the road map for making large purchases like a home.

Step 1—Estimate how much you need: The first question you need to answer is how much it would cost you to achieve that goal.

Step 1b—Consider other financial obligations you have at the moment. Sum your income and gather all your debt. What percentage does debt eat up from your income? Is debt taking up 10%, 30%, 50%?

Step 2—Reevaluate budget: Reviewing your budget is a step you cannot escape. Doing this gives you the broad picture of your situation and how planning for a house or other major purchases may affect your finances. Study your financial circumstances; will the amount of debt allow you to save enough money for a house? Do you need to sacrifice one aspect of your life to save more money? What is the cost of buying a house?

Step 3—Set up an account: You have two options. You could opt to save for the house using a regular savings account or invest your money into an investment account. Recall that the investment account works magic on your money, but bear in mind that it comes with risk. Seek the advice of a professional advisor before investing money.

Step 4—Consider economic activities: The economy has a huge impact on your goals in many ways. With housing, the prices of homes may increase and cripple your plans entirely. Knowing when to buy a house becomes essential. During inflation, the cost of goods increases and will continue to do so if the economy doesn't improve. If conditions aren't favorable now, are you willing to wait? What if it goes on forever?

Step 5—Figure out a payment plan: How do you intend to pay for the item or house? Cash or mortgage? If you're paying with cash, how long will you have to save? If you're paying with a mortgage, how long can you afford to back the debt? More importantly, how does this payment affect other financial goals?

STAYING AHEAD OF THE CURVE WITH LIFELONG LEARNING

The finance world is dynamic, and success tomorrow requires knowledge beyond today. To thrive in this ever-changed landscape, adopting a lifelong learning mindset is crucial. Lifelong learning means continuously learning and developing your skill set in all aspects of your life.

Undoubtedly, the more you educate yourself about your finances, the higher your quality of life. That's why lifelong learning should be an integral part of your journey. You don't have to spend all your savings to acquire finance skills. Finance educational materials are easily accessible.

Thankfully, your mobile phone serves as a link to the outside world. Even social media is a great source for information, if you know where to look.

 "Anyone who stops learning is old, at twenty or eighty. Anyone who keeps learning stays young."

— *HENRY FORD*

The question is, how can you embrace lifelong learning? Lifelong learning can be accessed in various forms. You could:

1. Enroll for formal finance classes, such as a one-year diploma program, Master of Business Administration, or other finance and accounting courses.
2. Consider self-directed learning, a path I've personally followed. While it may take longer and is less formal, nothing beats learning at your own pace. Check out:
3. Online learning platforms like EDx, Coursera, or Udemy offer a lot of financial self-taught courses you could take on.
4. Finance podcasts could be a great avenue to obtain quality finance updates.

5. If you can avoid getting trapped in the reels, YouTube is another university of free educational material.

Don't forget the saying, "Google is your friend!" What are you waiting for? Choose one avenue that you're most comfortable with. Commit to your personal growth and financial success by dedicating time each day or week to expand your knowledge. The opportunities out there are limitless, and the rewards are invaluable. Get started now!

END OF SESSION

In this chapter, we've explored the key aspects of growing up and preparing for the future. We talked about mastering skills that schools might not teach you, like acing job hunts, writing impressive job applications, and even building your own professional image.

Adulthood hits you with a wave of responsibilities. For example, one of your top priorities will be saving for rent. But life's journey doesn't stop there. As time goes on, you'll begin to prepare for relationships and maybe even marriage, each event coming with its own unique challenges. We discussed the importance of being open and honest in relationships and shared tips for planning your financial journey with a partner.

I also introduced the concept of financial reassessment as life progresses and goals change. The significance of starting to learn the workings of investment and retirement planning cannot be overstated. We also explored several types of investment vehicles and investment accounts you could use to fulfill your dreams.

Life is like a never-ending roller coaster ride with lots of twists and turns, and it's crucial that you're always prepared. As we navigate this roller coaster called "life," remember: Don't be a flickering candle in the wind. That's why I want to emphasize the significance of lifelong learning, which will empower you to navigate this exciting journey.

Your Input Matters!

This is your chance to make a huge difference, both to other teenagers seeking to understand money matters, and to me as I go forward to write future books. Your feedback is essential to the process – and this is your chance to leave it.

Simply by sharing your honest opinion of this book and a little about what you discovered here, you'll help new readers find the information they're looking for, as well as help to improve the resources I put out in the future.

LEAVE A REVIEW!

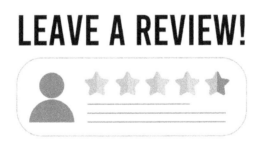

Thank you so much for your support. I wish you a lifetime of success and abundance.

Scan the QR code below to leave your review!

CONCLUSION

We have come to the end of this journey, and I must say, you've done a great job getting to this point. You deserve a superhero badge! The pages of this book have taught you the importance of money management, cultivating a positive money mindset, and the art of budgeting. You've discovered the power of saving and how to make your money work for you through investing.

In the chapters on credit and debt, you've learnt the formula for building a positive credit history and navigating the, sometimes tricky, world of borrowing. You've also explored the world of part-time jobs and entrepreneurship, equipped with the tools to polish your career, ace interviews, and even start your own business.

Don't forget, impulsive buying eats your finances faster than light travels from the sun to the earth. That's why smart shopping, combined with saving techniques, should become second nature to you.

You can now confidently approach the doors of college knowing you possess the knowledge of how to plan your finances during college and seek out financial aid when necessary. After that phase comes the career planning and job-hunting aspect of your life. It is inevitable and scary, but, like Thanos, you've collected all the infinity stones to help you conquer your job hunt and help you manage the financial responsibilities that come with adulthood.

You've also realized that you don't need to be a financial wizard to take control of your life. Simple, consistent actions can lead to extraordinary outcomes. Here's your chance to prove to yourself that you can be financially savvy, responsible, and free.

Your age is not a barrier; it's an advantage. Many successful individuals started their financial education at your age, and so can you. Dream big and never underestimate the endless possibilities and freedoms that financial literacy can bring to your life. Embrace this opportunity, and remember, the future is yours to shape!

So, march into the world with confidence; you have the super-powers of financial literacy at your disposal. The world awaits you!

Hopefully, I've helped you become a better you. But I need your help too; let me know how this book has improved your life. What areas of your financial life did I miss out on? Don't be shy to leave a review, and I'll be sure to check it out. Thank you!

REFERENCES

ADAMCZYK, B. (2023, March 24). *Around 85% Of Gen Zers Worry They Can't Afford One Month's Expenses if They Lose Their Job.* Fortune. https://fortune.com/2023/03/24/gen-zers-worried-about-covering-expenses-after-job-loss/

Akita, L. G. (n.d.). *A Quote by Lailah Gifty Akita.* Goodreads. https://www.goodreads.com/quotes/10565807-everything-is-in-excess-except-money-thereof-it-should-be

AZ Quotes. (n.d.). *Top 25 War Strategy Quotes (of 71).* A-Z Quotes. https://www.azquotes.com/quotes/topics/war-strategy.html

Burt, C. (2022, June 7). *Why 50% of Gen Z Students Say They See Less Value in College degrees.* University Business. https://universitybusiness.com/why-50-of-gen-z-students-say-they-see-less-value-in-college-degrees/

Cagnassola, M. E. (2022, August 28). *Here's How Much Debt the Average American Has in 2023.* Money. https://money.com/average-american-personal-debt-amount/#:

Career One Stop. (2020). *CareerOneStop-Find Scholarships.* Careeronestop. https://www.careeronestop.org/Toolkit/Training/find-scholarships.aspx

Cruze, R. (2021, December 14). *Impulse buying: Why We Do it And How To Stop.* Ramsey Solutions. https://www.ramseysolutions.com/budgeting/stop-impulse-buys

Davis, M. (2023, August 21). *26% of Americans Have Taken on Debt for New Tech.* LendingTree. https://www.lendingtree.com/debt-consolidation/tech-debt-survey/

Dickler, J. (2022, October 24). *63% Of Americans Are Living Paycheck to Paycheck — Including Nearly Half of Six-Figure Earners.* CNBC. https://www.cnbc.com/2022/10/24/more-americans-live-paycheck-to-paycheck-as-inflation-outpaces-income.html

Dixon, A. (2019, March 14). *Survey: 1 In 5 Working Americans Aren't Saving Anything At All | Bankrate.com.* Bankrate. https://www.bankrate.com/banking/savings/financial-security-march-2019/

Dow, N. (2021, June 15). *These Budgeting Statistics Show Most of Us Don't Track*

Our Spending. Www.thepennyhoarder.com. https://www.thepenny hoarder.com/budgeting/budgeting-statistics/

Elkins, K. (2019, May 10). *College is Almost Always Worth The Cost, Says Wealth Manager—Here's Why.* CNBC. https://www.cnbc.com/2019/05/10/wealth-manager-why-college-is-almost-always-worth-the-cost.html

Equifax. (n.d.). *Equifax.* Equifax.com. https://www.equifax.com/personal/education/credit/score/what-is-a-credit-score/

Fay, B. (2022, February 23). *Consumer Debt Statistics & Demographics in America.* Debt.org. https://www.debt.org/faqs/americans-in-debt/demographics/

Felton, K. (2023, February 28).Expectations vs Reality: The Cost of Raising Children in 2023. Education Matters Magazine. https://www.educationmatters mag.com.au/expectations-vs-reality-the-cost-of-raising-children-in-2023/

45+ Eye-Opening Superpowers Quotes That Will Inspire Your Inner Self. Quotlr - Best Quotes from Top Authors. Accessed December 23, 2023. https://quotlr.com/quotes-about-superpowers.

Gayle, L. (2021, May 21). *Teen Cashed In £200,000 From £200 After Teaching Himself About Stocks.* Mail Online. https://www.dailymail.co.uk/femail/arti cle-9600731/Teenager-cashed-200-000-200-teaching-invest-stocks.html

Gilbert, N. (2020, February 28). *25 Statistics & Reasons Not to Go to College: 2020 Data & Analysis.* Financesonline. https://financesonline.com/reasons-not-to-go-to-college-statistics/

Goodreads. (n.d.). *Savings Quotes (24 quotes).* Www.goodreads.com. https://www.goodreads.com/quotes/tag/savings

Hanson, M. (2023, April 1). *Student Loan Debt Statistics [2020]: Average + Total Debt.* EducationData. https://educationdata.org/student-loan-debt-statistics

Institute for Divorce and Financial Analysts. (n.d.). *Why People Divorce and What Are the Reasons for Divorce?* Institutedfa.com. https://institutedfa.com/Leading-Causes-Divorce

iwillshampooyouitsok. (2021, September 5).I Feel So Lost and Stupid for Having No Retirement Savings.. Reddit. https://www.reddit.com/r/DaveR amsey/comments/pi3r7h/i_feel_so_lost_and_stupid_for_having_no/

Kagan, J. (2023, May 17). *Credit Score.* Investopedia. https://www.investopedia.com/terms/c/credit_score.asp

Lipson, S. (2020, May 18). *Insurance "Saved the Day" Stories.* Www.linkedin.com.

https://www.linkedin.com/pulse/insurance-saved-day-stories-shane-lipson/

Liska, C. (2017, December 8). *Measuring Coaching ROI*. Main. https://www.td.org/insights/measuring-coaching-roi

Lodge, M. (2019, October 25). *10 Successful Young Entrepreneurs*. Investopedia. https://www.investopedia.com/10-successful-young-entrepreneurs-4773310

Massa, A., & Witzig, J. (2023, July 3). Musk, Zuckerberg Lead a $852 Billion Surge Among World's Richest People. *Bloomberg*. https://www.bloomberg.com/news/articles/2023-07-03/musk-zuckerberg-lead-852-billion-wealth-surge-among-world-rich#:

Mickle, T. (2023, September 11). *As Smartphone Industry Sputters, the iPhone Expands Its Dominance*. The New York Times. https://www.nytimes.com/2023/09/11/technology/apple-iphone-17.html

Quinn, R. (2013, July 15). *Cow Crashes Through Roof, Kills Sleeping Man*. USA TODAY. https://www.usatoday.com/story/news/world/2013/07/15/newser-cow-kills-man/2517321/

Reinicke, C. (2022, March 3). *60% Of Teens Want to Launch Their Own Businesses Instead of Working Regular Jobs*. CNBC. https://www.cnbc.com/2022/03/03/60percent-of-teens-want-to-launch-businesses-instead-of-working-regular-jobs.html

Rogers, S. (2022, September 16). *Relationships & Finance: How Do They Interact? (Survey)*. FinMasters. https://finmasters.com/relationships-and-finance/#gref

Saint-Exupéry, A. de . (n.d.). *Best 30 Planning Quotes You Need to Know About: Ultimate Guide*. Instagantt. https://instagantt.com/project-management/best-30-planning-quotes

Sharma, R. S. (n.d.). *A quote from The Monk Who Sold His Ferrari*. Goodreads. https://www.goodreads.com/quotes/493669-investing-in-yourself-is-the-best-investment-you-will-ever

Statistica. (2020, March). *U.S. Monthly Apartment Rent 2016-2020*. Statista. https://www.statista.com/statistics/1063502/average-monthly-apartment-rent-usa/

Tenny, J. (2022, April 4). *Survey Finds 93% of Teens Believe Financial Knowledge and Skills Are Needed to Achieve Their Life Goals*. Www.businesswire.com. https://www.businesswire.com/news/home/20220404005339/en/Survey-

Finds-93-of-Teens-Believe-Financial-Knowledge-and-Skills-Are-Needed-to-Achieve-Their-Life-Goals

The Investopedia Team. (2023, May 17). *What Is a Credit Score? Definition, Factors, and Ways to Raise It*. Investopedia. https://www.investopedia.com/terms/c/credit_score.asp#:

Understood. (n.d.). *How to Help Your Teen Balance School and a New Job*. Www.understood.org. https://www.understood.org/en/articles/balancing-school-and-a-new-job

Vanguard. (n.d.). *Roth IRA: What is a Roth IRA? Vanguard*. https://investor.vanguard.com/accounts-plans/iras/roth-ira#:

Ward, P. (2019, August 20). *63% of Americans Think Personal Finance Education Belongs in School*. Intuit Credit Karma. https://www.creditkarma.com/insights/i/personal-finance-in-schools-survey#:

Ycharts. (n.d.). *US Unemployment Rate*. Ycharts.com. https://ycharts.com/indicators/us_unemployment_rate#: